About this book

This book has been written specifically for Year 9 of the Framework for Teaching Mathematics. It is aimed at students who are following the Year 9 teaching programme from the Framework and leads to the 5–7 tier of entry in the NC tests.

The authors are experienced teachers and maths consultants who have been incorporating the Framework approaches into their teaching for many years and so are well qualified to help you successfully meet the Framework objectives.

The book is made up of units based on the sample medium term plans which complement the Framework document, thus maintaining the required pitch, pace and progression.

The units are:

A1	Sequences	1–8
A2	Functions and graphs	9–16
N1	Proportional reasoning	17–36
A3	Solving equations	37–50
S1	Geometrical reasoning and construction	51–70
D1	Handling data	71–84
S2	Perimeter, area and volume	85–98
N2	Number calculations	99–116
A4	Graphs of functions	117–136
D2	Probability	137–146
S3	Transformations and congruence	147–160
P1	Problem solving and revision	161–174
A5	Equations, functions and formulae	175–188
D3	Statistical reports	189–202
S4	Applying geometrical reasoning	203–216
D4	Theoretical and experimental probability	217–230
B1	Trigonometry and Pythagoras' theorem	231–244

The last five units in this book are designed to consolidate KS3 work and bridge to KS4 work.

Each unit comprises double page spreads that should take a lesson to teach. These are shown on the full contents list.

Problem solving is integrated throughout the material as suggested in the Framework.

How to use this book

This book is made up of units of work which are colour coded into: Algebra (Blue), Data (Pink), Number (Orange), Shape, space and measures (Green), Problem solving (Light Green) and a Bridging unit (Red).

Each unit of work starts with an overview of the content of the unit, as specified in the Framework document, so that you know exactly what you are expected to learn.

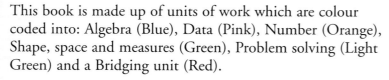

This unit will show you how to:

▶▶ Find the inverse of a linear function.

▶▶ Construct functions arising from real-life problems and plot their corresponding graphs.

▶▶ Interpret graphs arising from real situations.

▶▶ Represent problems and synthesise information in algebraic or graphical form.

The first page of a unit also highlights the skills and facts you should already know and provides Check in questions to help you revise before you start so that you are ready to apply the knowledge later in the unit:

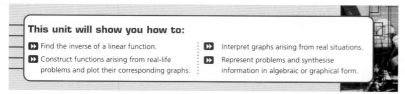

Before you start

You should know how to ...

1 Express functions in symbols.

2 Draw a mapping diagram for a simple function.

Check in

1 Find the output of this function machine:

$$n \longrightarrow \boxed{\times 2} \longrightarrow \boxed{-1} \longrightarrow ?$$

2 Copy and complete this mapping diagram for the function $x \rightarrow 2x - 1$

Inside each unit, the content develops in double page spreads which all follow the same structure.

The spreads start with a list of the learning outcomes and a summary of the keywords:

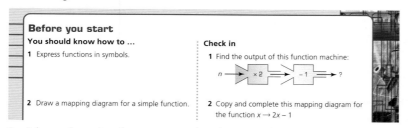

This spread will recap how to:

▶▶ Express linear functions in symbols.

▶▶ Find the inverse of a linear function.

This sequence has two extra counters in each pattern:

KEYWORDS

Function   Mapping
General term Sequence
Operation   Term
Inverse function
Function machine

The keywords are summarised and defined in a Glossary at the end of the book so you can always check what they mean.

Key information is highlighted in the text so you can see the facts you need to learn.

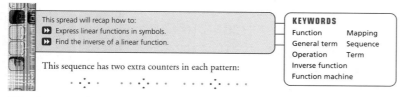

▶ You can divide an integer by any fraction.

Examples showing the key skills and techniques you need to develop are shown in boxes. Also hint boxes show tips and reminders you may find useful:

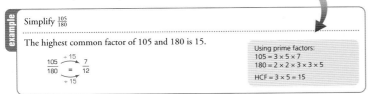

Simplify $\frac{105}{180}$

The highest common factor of 105 and 180 is 15.

$\frac{105}{180} \xrightarrow[\div 15]{\div 15} = \frac{7}{12}$

Using prime factors:
$105 = 3 \times 5 \times 7$
$180 = 2 \times 2 \times 3 \times 3 \times 5$
$HCF = 3 \times 5 = 15$

Remember:
Multiplying by $\frac{1}{7}$ is the same as dividing by 7.

Each exercise is carefully graded, set at three levels of difficulty:

▸▸ The first few questions provide lead-in questions, revising previous learning.
▸▸ The questions in the middle of the exercise provide the main focus of the material.
▸▸ The last few questions are challenging questions that provide a link to further learning objectives.

At the end of each unit is a summary page so that you can revise the learning of the unit before moving on.

Check out questions are provided to help you check your understanding of the key concepts covered and your ability to apply the key techniques. These are all based on actual Key Stage 3 paper questions so they give you practice at the standard required in your examination.

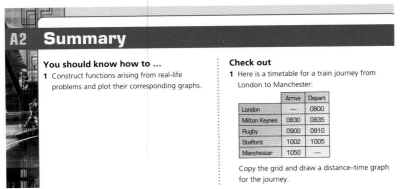

A2 Summary

You should know how to ...

1 Construct functions arising from real-life problems and plot their corresponding graphs.

Check out

1 Here is a timetable for a train journey from London to Manchester:

	Arrive	Depart
London	—	0800
Milton Keynes	0830	0835
Rugby	0900	0910
Stafford	1002	1005
Manchester	1050	—

Copy the grid and draw a distance–time graph for the journey.

The answers to the Check in and Check out questions are produced at the end of the book so that you can check your own progress and identify any areas that need work.

Contents

A1 Sequences   1–8

A1.1	Revising sequences	2
A1.2	Using practical contexts	4
A1.3	Quadratic sequences	6
	Summary	8

A2 Functions and graphs   9–16

A2.1	Revising functions	10
A2.2	Distance–time graphs	12
A2.3	Interpreting graphs	14
	Summary	16

N1 Proportional reasoning   17–36

N1.1	Adding and subtracting fractions	18
N1.2	Multiplying by fractions	20
N1.3	Dividing by fractions	22
N1.4	Proportions of amounts	24
N1.5	Comparing proportions	26
N1.6	Percentage change	28
N1.7	Using ratio	30
N1.8	Proportionality	32
N1.9	Order of operations	34
	Summary	36

A3 Solving equations   37–50

A3.1	Using letter symbols	38
A3.2	Balancing linear equations	40
A3.3	Solving non-linear equations	42
A3.4	Solving proportional problems	44
A3.5	Trial and improvement	46
A3.6	Best estimate	48
	Summary	50

S1 Geometrical reasoning and construction — 51–70

S1.1	Geometrical language	52
S1.2	Angles in polygons	54
S1.3	Angles in regular polygons	56
S1.4	Lines and angles	58
S1.5	Solving problems in shapes	60
S1.6	Circle properties	62
S1.7	Constructing right-angled triangles	64
S1.8	Drawing loci	66
S1.9	More loci	68
	Summary	70

D1 Handling data — 71–84

D1.1	The handling data cycle	72
D1.2	Calculating statistics	74
D1.3	Representing data	76
D1.4	Interpreting graphs and charts	78
D1.5	Comparing distributions	80
D1.6	Misleading statistics	82
	Summary	84

S2 Perimeter, area and volume — 85–98

S2.1	Measuring area	86
S2.2	Circumference of a circle	88
S2.3	Area of a circle	90
S2.4	Circle problems	92
S2.5	Measuring volume	94
S2.6	Prisms	96
	Summary	98

N2 Number calculations   99–116
N2.1   Powers and roots   100
N2.2   Powers of 10   102
N2.3   Rounding   104
N2.4   Addition and subtraction   106
N2.5   Mental multiplication and division   108
N2.6   Written multiplication   110
N2.7   Written division   112
N2.8   Calculator methods   114
       Summary   116

A4 Graphs of functions   117–136
A4.1   Finding factors   118
A4.2   Using prime factors   120
A4.3   Index notation   122
A4.4   Solving problems involving indices   124
A4.5   Plotting linear graphs   126
A4.6   Finding the gradient   128
A4.7   Implicit functions   130
A4.8   Real-life graphs   132
A4.9   Interpreting distance–time graphs   134
       Summary   136

D2 Probability   137–146
D2.1   Probability of an event   138
D2.2   Mutually exclusive outcomes   140
D2.3   Calculating probabilities   142
D2.4   Experimental probability   144
       Summary   146

S3 Transformations and congruence   147–160
S3.1   Transformations   148
S3.2   Describing transformations   150
S3.3   Combining transformations   152
S3.4   Planes of symmetry   154
S3.5   Enlargement and similarity   156
S3.6   Scale and proportion   158
       Summary   160

P1 Problem solving and revision   161–174

P1.1	Making sense of the problem	162
P1.2	Answering the question	164
P1.3	Choosing a strategy	166
P1.4	Proportional reasoning	168
P1.5	Geometrical reasoning	170
P1.6	Checking results	172
	Summary	174

A5 Equations, functions and formulae   175–188

A5.1	Equations, functions, formulae and identities	176
A5.2	Using brackets	178
A5.3	Solving equations	180
A5.4	Algebraic fractions	182
A5.5	Transforming formulae	184
A5.6	Solving equations using graphs	186
	Summary	188

D3 Statistical reports   189–202

D3.1	Gathering data	190
D3.2	Organising data	192
D3.3	Scatter graphs and correlation	194
D3.4	Choosing the correct statistics	196
D3.5	The shape of distribution	198
D3.6	Statistical reports	200
	Summary	202

S4 Applying geometrical reasoning   203–216

S4.1	Properties of 2-D shapes	204
S4.2	Properties of angles	206
S4.3	2-D representation of 3-D objects	208
S4.4	Scale drawings and ratio	210
S4.5	Maps and ratio	212
S4.6	Surface area and volume of a prism	214
	Summary	216

**D4 Theoretical and experimental
 probability**   217–230
  D4.1 Combined events   218
  D4.2 Mutually exclusive outcomes   220
  D4.3 Tree diagrams   222
  D4.4 More tree diagrams   224
  D4.5 Comparing theoretical and
        experimental probability   226
  D4.6 Probability experiments   228
        Summary   230

B1 Trigonometry and Pythagoras' theorem   231–244
  B1.1 Triangles in a circle   232
  B1.2 Sines and cosines   234
  B1.3 Opposite and adjacent   236
  B1.4 Tangents   238
  B1.5 Pythagoras' theorem   240
  B1.6 Triangle problems   242
        Summary   244

Glossary   245
Answers   265
Index   274

1 Sequences

This unit will show you how to:

▶▶ Generate terms of a sequence using-term-to-term and position-to-term definitions of the sequence.

▶▶ Generate sequences from practical contexts.

▶▶ Write an expression to describe the nth term of an arithmetic sequence.

▶▶ Generate terms of a quadratic sequence.

▶▶ Solve increasingly demanding problems and evaluate solutions.

▶▶ Present a concise, reasoned argument using symbols and explanatory text.

▶▶ Represent problems and synthesise information in algebraic form.

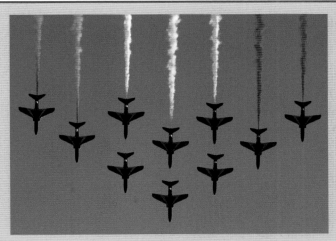

You find sequences in surprising places!

Before you start

You should know how to ...

1 Find missing terms in a sequence.

2 Substitute values into algebraic expressions.

Check in

1 Find the missing terms:

    a 1, 4, 7, 10, __, __

    b ⁻3, 0, __, __, 9, 12

    c ⁻1, ⁻5, __, __, ⁻17, ⁻21

    d 2.3, 2.5, __, __, __, 3.3

2 Substitute $n = 2$ into each of these expressions:

    a $3n + 1$ b $2 + 6n$

    c $6 - n$ d $9 - 4n$

This spread will show you how to:

▶▶ Generate and describe integer sequences.
▶▶ Generate terms of a simple sequence given a rule.
▶▶ Express simple functions using symbols.

KEYWORDS

Sequence   T(n)
Term   Generate
General term
Position-to-term rule
Term-to-term rule

To generate or describe a sequence you need:

▶ a term-to-term rule and a term, or
▶ a position-to-term rule.

A term-to-term rule links a term with the next term.

A position-to-term rule links a term to its position in the sequence.

Using a term-to-term rule

A sequence has:

▶ A first term of 5.
▶ You write T(1) = 5.
▶ A term-to-term rule of subtract 4.
▶ You subtract 4 to find the next term.

The first five terms are:

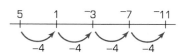

Using a position-to-term rule

The general term of a sequence is: $T(n) = 5n + 4$

To find T(1), you substitute $n = 1$: $T(1) = 5 \times 1 + 4 = 9$
To find T(4), you substitute $n = 4$: $T(4) = 5 \times 4 + 4 = 24$

n is the position of the term in the sequence.

---

example

Here is a sequence of patterns.
Each pattern has one black tile in the centre.
Each new pattern has more tiles.

  a   How many more tiles are added each time?
  b   How many tiles does the tenth pattern, T(10), have?
  c   Which of these functions gives the general term for the sequence?

      $4n + 3$ $3n + 4$ $n + 3$ $3n + 1$

....

  a   3 tiles are added each time – one on each arm.
  b   31 – that is 10 on each arm and 1 in the centre.
  c   Three tiles are added each time. The sequence is $3n +$ something: $3n + 1$

Exercise A1.1

1 Here is a rule to find the next term of a sequence: Add □.
Choose a first term for the sequence, T(1), and a number to go in the box so that your rule will generate all the terms of the sequence:
  a multiples of 5   **b** even numbers
  c odd numbers
  d all numbers ending in the same digit.

2 Write down the first five terms of a sequence when the nth term or T(n) is:
  a $3n - 2$ **b** $8n + 3$
  c $18 - 4n$ **d** $2n - \frac{1}{2}$
  e $0.5n + 1.5$ **f** $3.5 - 0.5n$

3 Here is a number chain:
$$1 \rightarrow 3 \rightarrow 5 \rightarrow 7 \rightarrow 9 \rightarrow$$
The rule is 'The first term is 1, add on 2 each time'.
  a A different number chain is:
    $$1 \rightarrow 3 \rightarrow 9 \rightarrow 27 \rightarrow 81 \rightarrow 243 \rightarrow$$
    What is the rule?
  b A chain starts like this:
    $$1 \rightarrow 5 \rightarrow \_\_ \rightarrow \_\_ \rightarrow \_\_ \rightarrow$$
    Show three different ways to continue this number chain.
    For each chain, write down the next three numbers and the rule you are using.

4 For each sequence, write down the first five terms.

	1st Term	Term-to-term rule
a	7	add 6
b	4	subtract 3
c	6	subtract 4
d	⁻3	add 6

5 Match each of these general terms with a sequence from question 4.
$$10 - 4n, \ 6n - 9, \ 7 - 3n, \ 6n + 1$$

6 Charmaine is making a series of patterns with black and white tiles.

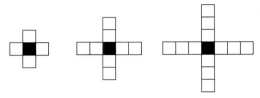

Charmaine says: 'The rule for finding the number of tiles in pattern number n is: number of tiles T(n) = $4n + 1$'.
  a The '1' in Charmaine's rule represents the black tile.
    What does the $4n$ represent?
  b Charmaine wants to make pattern number 12.
    How many black tiles and how many white tiles does she need?
  c Charmaine use 61 tiles altogether to make a pattern.
    What is the number of the pattern she makes?
  d Charmaine has 100 black tiles and 100 white tiles.
    What is the pattern number of the biggest pattern Charmaine can make?

7 Here are the 2nd and 3rd patterns in a series.

  a Draw the first pattern in the series.

Jack says the rule for finding the number of lines on pattern number n is: $4n + 4$.
Jill says the rule is: $2n + 2$.
They each explain what they mean by their rule and the teacher marks both correct.

  b Explain Jack's rule and Jill's rule so that they are both correct.

Using practical contexts

This spread will show you how to:

▶▶ Generate sequences from practical contexts.

▶▶ Write an expression to describe the *n*th term of an arithmetic sequence.

KEYWORDS

Sequence

General term

T(*n*)

This sequence is made from patterns of counters:

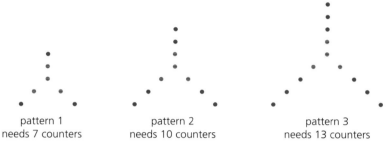

pattern 1
needs 7 counters

pattern 2
needs 10 counters

pattern 3
needs 13 counters

The core of each pattern is made up of four red counters.

There are three extra counters in each pattern – one on each arm.

The general term for the sequence is:

$$T(n) = 3n + 4$$

The extra 3 for each diagram   The core of 4 red counters

Always check your general term works by substituting:

$$T(1) = 3 \times 1 + 4 = 7$$
$$T(2) = 3 \times 2 + 4 = 10$$
$$T(3) = 3 \times 3 + 4 = 13$$

---

example

Here is a sequence of stick patterns:

a How many extra sticks are there in each pattern?

b Explain how you know that the general term for the pattern is $T(n) = 4n + 5$

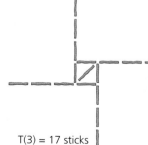

T(1) = 9 sticks   T(2) = 13 sticks   T(3) = 17 sticks

a There are four extra sticks in each pattern – one on each arm.

b The extra four sticks each time gives the $4n$.

There are five sticks in the central group, which gives the +5.

Exercise A1.2

1 Counter pattern match

Match each counter pattern with its position-to-term rule and its general term T(n).

Pattern		Rule		T(n)
a	**i**	T(1) = 7 T(2) = 9 T(5) = 15	**I**	T(n) = $4n + 3$
b	**ii**	T(1) = 7 T(2) = 11 T(5) = 23	**II**	T(n) = $2n + 5$
c	**iii**	T(1) = 8 T(2) = 11 T(5) = 20	**III**	T(n) = $3n + 5$

2 These huts have been made with matches.

T(1) = 6   T(2) = 11   T(3) = 16 matches

A rule to find how many matches are needed is T(n) = $5n + 1$.

a Use this rule to find how many matches are needed for 12 huts.

b I use 81 matches to make some huts. How many huts do I make?

c Different huts are made with matches:

T(1) = 5   T(2) = 9   T(3) = 13

Which of these rules shows how many matches are needed?

T(n) = $n + 4$ T(n) = $5n + 4$

T(n) = $4n + 5$ T(n) = $4n + 1$

T(n) = $5n - 2$

Explain why you have chosen this rule.

3 Here is a sequence of tile patterns.

Pattern 1	Pattern 2	Pattern 3
T(1)	T(2)	T(3)

a Look at the patterns carefully and complete the table. Check the pattern number carefully.

Pattern number	Number of blue tiles	Number of white tiles
T(1)		
T(3)	4	
T(5)		
T(16)		
T(n)		

b Look at the numbers of blue and white tiles for the general term T(n) and write an expression to show the total number of tiles in the pattern. Simplify your expression.

This spread will show you how to:
- ▶▶ Generate terms of a quadratic sequence.
- ▶▶ Write an expression to describe the *n*th term of an arithmetic sequence.

KEYWORDS

General term   T(*n*)
Linear sequence   Difference
Position-to-term rule
Term-to-term rule

A linear sequence changes by the same amount each time:

2, 5, 8, 11, 14, …	6.5, 5, 3.5, 2, 0.5, …
The first term is 2.	The first term is 6.5.
The term-to-term rule is $+ 3$.	The term-to-term rule is -1.5.
The position-to-term rule (*n*th term) is: $T(n) = 3n - 1$	The position-to-term rule (*n*th term) is: $T(n) = 8 - 1.5n$

A linear sequence has a constant difference between terms:

$$T(1) = 3 \qquad T(2) = 8 \qquad T(3) = 13 \qquad T(4) = 18$$

is a linear sequence with a constant difference of 5.

A quadratic sequence changes by an increasing or decreasing amount:

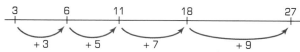

The first term is 3.
The term-to-term rule is 'add the next odd number'.

The general term for this sequence is $T(n) = n^2 + 2$.

You can use the general term to find other terms:

$$T(8) = 8^2 + 2$$
$$= 64 + 2 = 66$$

The differences 3, 5, 7, 9 form a linear sequence.

The n^2 term shows the sequence is quadratic.

example

Here is a sequence of patterns made from counters.

- **a** Describe in words how the number of counters increases.
- **b** Find T(1), T(2), T(3), T(5) and T(10).

- **a** The first pattern has 1 counter. The patterns increase by 3, 5, 7 and so on.
- **b** $T(1) = 1$, $T(2) = 4$, $T(3) = 9$, $T(5) = 25$, $T(10) = 100$

Exercise A1.3

1 For each of these sequences, you are given the first three terms and the sixth term. Work out T(4) and T(5) and explain the sequence in words.

  a T(1) = 2, T(2) = 5, T(3) = 10, T(6) = 37

  b T(1) = 0, T(2) = 1, T(3) = 3, T(6) = 15

  c T(1) = 2, T(2) = 5, T(3) = 11, T(6) = 47

  d T(1) = 25, T(2) = 16, T(3) = 9, T(6) = 0

2 Write down the first five terms in each of these sequences.

  a The first term is 2 and the term-to-term differences increase by the even numbers 2, 4, 6, 8, ...

  b T(1) = 24 and the term-to-term differences decrease by the odd numbers 3, 5, 7, ...

  c $T(n) = n^2 + 7$

  d $T(n) = n(n + 1)$

  e $T(n) = 2n^2 - 3$

3 This sequence pattern matches one of the five sequences in question 2.

Which one does it match?

4 Find the general term for each of these linear sequences:

  a 21, 27, 33, 39, 45, ...

  b 54, 62, 70, 78, 86, ...

  c 68, 61, 54, 47, 40, ...

  d 2.3, 2.5, 2.7, 2.9, 3.1, ...

  e ⁻5, ⁻14, ⁻23, ⁻32, ⁻41, ...

5 Here is a sequence of patterns made from counters.

pattern 1   pattern 2   pattern 3   pattern 4

  a Copy and complete the table.

Pattern	1	2	3	4	5	10
Number of counters		3				

  b Describe the term-to-term rule.

  c This sequence of numbers is known as the set of triangular numbers and $T(n) = \frac{1}{2}n(n + 1)$.
Use this expression (formula) to find T(8) and T(15).

6 Here is a sequence of square patterns.

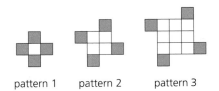

pattern 1   pattern 2   pattern 3

  a How many white tiles will be used in each of the first five patterns?

  b How many white tiles and how many red tiles will be used in pattern number 8?

  c How many white tiles and how many red tiles will be used in pattern number 12?

  d How many white tiles and how many red tiles will be used in pattern number n?

  e Write down an equation to show the total number of tiles used in T(n):
T(n) =

You should know how to ...

1 Generate terms of a sequence using term-to-term and position-to-term definitions of the sequence.

2 Write an expression to describe the nth term of an arithmetic sequence.

3 Present a concise, reasoned argument, using symbols and explanatory text.

Check out

1 Work out the first, third and sixth terms for each of these sequences.

   a $2n + 7$

   b $5n - 8$

   c $10 - 2n$

   d $3n - 20$

   e $3 - 2.5n$

2 For each of these linear sequences, explain the sequence in words and find the general term $T(n)$.

   a   4, 7, 10, 13, 16, ...

   b   9, 11, 13, 15, 17, ...

   c   20, 18, 16, 14, 12, ...

   d   8, 13, 18, 23, 28, ...

3 For this sequence of lines:

   a   Write the first five terms of the sequence of lines.

   b   Work out $T(10)$ and $T(25)$.

   c   Explain in your own words how you worked out $T(25)$.

   d   Find $T(n)$, where n is the pattern number.

This unit will show you how to:

▶▶ Find the inverse of a linear function.

▶▶ Construct functions arising from real-life problems and plot their corresponding graphs.

▶▶ Interpret graphs arising from real situations.

▶▶ Represent problems and synthesise information in algebraic or graphical form.

A computer can process data – you input values and formulae and it outputs the result.

Before you start

You should know how to ...

1 Express functions in symbols.

2 Draw a mapping diagram for a simple function.

3 Plot points on a coordinate grid.

Check in

1 Find the output of this function machine:

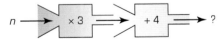

$$n \longrightarrow \boxed{\times 3} \Longrightarrow \boxed{+ 4} \longrightarrow ?$$

2 Copy and complete this mapping diagram for the function $x \to 2x - 1$.

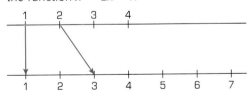

3 Plot the points $(^-2, 1)$ $(0, 3)$ $(4, 1)$ $(0, ^-1)$.

Join them in order.

What shape have you drawn?

This spread will show you how to:
▶▶ Express linear functions in symbols.
▶▶ Find the inverse of a linear function.

KEYWORDS
Function   Mapping
General term   Sequence
Operation   Term
Inverse function
Function machine

This sequence has two extra counters in each pattern:

pattern 1   pattern 2   pattern 3
6 counters   8 counters   10 counters

You can write the general term of the sequence as a function: $T(n) = 2n + 4$

The function maps the position onto the term: $n \longrightarrow 2n + 4$

You can write the function in a function machine:

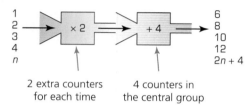

2 extra counters   4 counters in
for each time   the central group

You can work backwards to find the position of a term:
If there are 12 counters in a pattern ...

Remember these inverse
operations:
▶ × and ÷
▶ + and −

$4 \longleftarrow \div 2 \longleftarrow -4 \longleftarrow 12$ ... it must be the 4th pattern.

When you work backwards you find the inverse:

The **inverse function** is $n \longrightarrow \dfrac{(n-4)}{2}$

You can show a function and its inverse on a mapping diagram.

example

a Find the inverse of the function: $x \longrightarrow 2x - 1$
b Show the function and its inverse on a mapping diagram.

....

a Imagine the function machine:
The function is ×2 then −1.
The inverse is +1 then ÷2.
The inverse function is:
$$x \longrightarrow \frac{(x+1)}{2}$$

b The mapping diagram is:

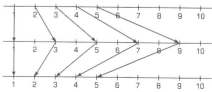

Exercise A2.1

1 Copy and complete these function machines.

a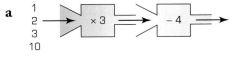

b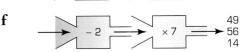

c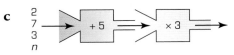

d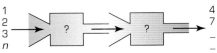

e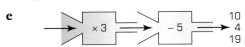

f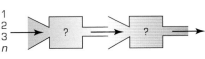

2 For each of these sequence patterns, copy and complete the function machine.

a

b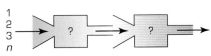

c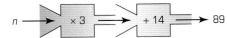

3 Use the inverse method to find the missing input values.

a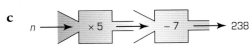

b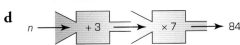

c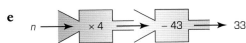

d

e

f

4 Find the inverse of these linear functions. The first one is done for you.

a $n \rightarrow 6n - 8$ inverse is $n \rightarrow \dfrac{(n+8)}{6}$

b $n \rightarrow 3n + 14$ **c** $n \rightarrow 5n - 7$ **d** $n \rightarrow 7(n+3)$ **e** $n \rightarrow \dfrac{n}{5} + 12$

5 Challenge

For the function $x \rightarrow 4 - x$:

a Copy and complete this function machine.

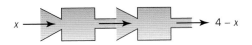

b Find the inverse function.

c Simplify your answer. What do you notice?

Distance–time graphs

This spread will show you how to:
> ▶ Construct functions arising from real-life problems and plot their corresponding graphs.
> ▶ Interpret graphs arising from real situations.

KEYWORDS
Graph   Axes
Time axis   Distance axis
Distance–time graph

A pop star travels from London to Carlisle and back for a concert. She uses a helicopter. This distance–time graph shows her journey.

The distance axis measures the distance away from London.

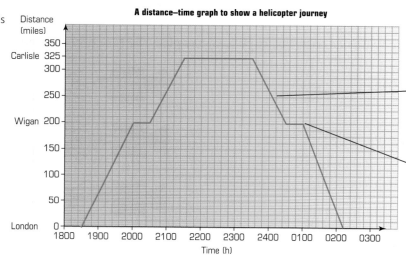

A distance–time graph to show a helicopter journey

This stage shows the helicopter moving back towards London.

When the graph is horizontal there is no movement – the helicopter has stopped to refuel.

The time axis shows the time taken for each stage.

You can find the average speed for any part of her journey:

> ▶ Average speed $= \dfrac{\text{distance}}{\text{time}}$

example

Use the graph for the helicopter journey to answer these questions.
  a How long was the pop star in Carlisle?
  b What was her average speed from London to Wigan?

....

  a The helicopter landed at 2130 and left at 2330. She was there for 2 hours.

  b London to Wigan took 1.5 hours to travel 200 miles. Speed $= \dfrac{200 \text{ miles}}{1.5 \text{ h}} = 133\frac{1}{3}$ miles per hour

To plot a distance–time graph:
▶ Choose suitable scales for the axes.
▶ Label the axes and give the graph a title.
▶ Plot each stage of the journey in turn.

Exercise A2.2

1 Jon won a holiday travelling to New York by the QM2 and returning by jet.
  a How many nights did he spend on board the QM2?
  b The return journey from New York by jet (2490 miles) took five hours. What was the average speed?
  c Write a paragraph to describe Jon's travels.

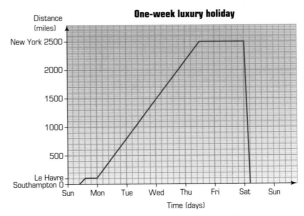

One-week luxury holiday

2 This graph shows an aeroplane travelling from London to Majorca via Barcelona and one flying directly to Majorca.
  a What is the first aeroplane's average speed from London to Barcelona?
  b How can you tell from the graph that the aeroplane's average speed from London to Barcelona is greater than from Barcelona to Majorca?
  c At what time are the two planes the same distance from London?

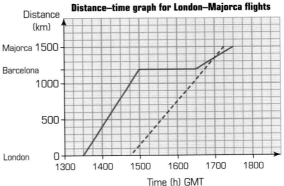

Distance–time graph for London–Majorca flights

3 On squared paper, copy this grid.
  a Here is part of the London Express timetable:

depart Edinburgh	0600
arrive York	0800
depart York	0845
arrive London	1130

On your grid, show the train's journey from Edinburgh to London.

  b The 'Early Riser' train follows this timetable:

London	depart	0630
Peterborough	arrive	0730
	depart	0745
York	arrive	0945
	depart	1000
Newcastle	arrive	1100
	depart	1115
Edinburgh	arrive	1215

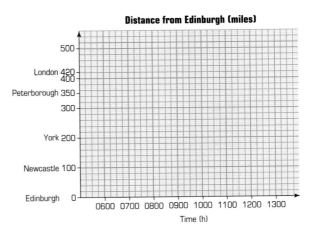

Distance from Edinburgh (miles)

Show this journey on the same graph.
  c At what time are the two trains the same distance from Edinburgh?
     How far are the trains from Edinburgh at this time?
  d When is the London Express travelling fastest?
     What is its average speed for this part of the journey?

13

Interpreting graphs

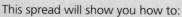

This spread will show you how to:

▶▶ Construct functions arising from real-life problems and plot their corresponding graphs.

▶▶ Interpret graphs arising from real situations.

KEYWORDS

Graph

Interpret

Distance–time graph

To cook frozen peas, Jamie boils some water, adds the peas and brings the water back to the boil.

This graph shows the changes in the temperature of the water:

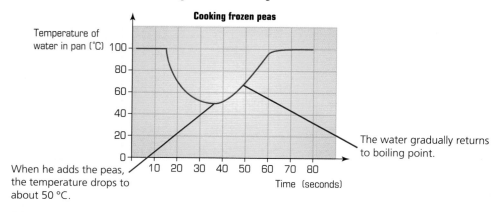

When he adds the peas, the temperature drops to about 50 °C.

The water gradually returns to boiling point.

You can use graphs to illustrate many real-life situations.

---

example

The table gives the official lighting-up times for street lamps in Birmingham.

a Draw a graph to display this information. Plot all the points and join them in a smooth curve.

b What are the earliest and latest lighting-up times?

c When was the lighting-up time at 1800?

All times in GMT

Date	Time	Date	Time
20th January	1700	20th July	2045
20th February	1745	20th August	2000
20th March	1840	20th September	1850
20th April	1940	20th October	1740
20th May	2025	20th November	1650
20th June	2050	20th December	1630

a The graph will look like this:

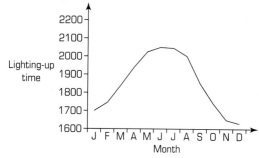

b The earliest lighting-up time is in December. The latest is in June.

c The lighting-up time was 1800 in February and October.

Exercise A2.3

1 This is a simplified graph showing the depth of water in a bath.

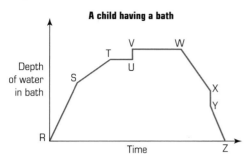

A child having a bath

Between R and T water fills the bath.
a What could have happened at point S?
b At what point did the child get into the bath?
c At what point did the child get out of the bath?
d In a short paragraph, describe the story of the child having a bath.

2 The graph shows the heights of two rockets during their flights at a firework display.

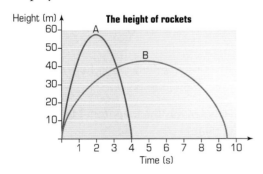

The height of rockets

a Rocket A flew for four seconds. What was the total flight time for rocket B?
b Estimate how much higher rocket A reached than rocket B.
c Estimate the time after the start when the two rockets were at the same height.
d Estimate how much longer rocket B was more than 30 m above the ground than rocket A.

3 Here are four containers.

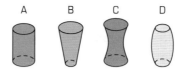

Water is poured at a constant rate into the containers.
The graphs show the depth of water as the containers fill up.

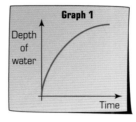

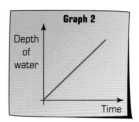

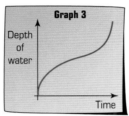

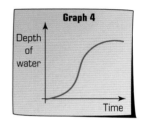

a Which graph matches each container?
b A fifth container looks like this: Sketch a graph to show the depth of water over time as the container fills up.

4 The highway code gives these minimum stopping distances for cars travelling at different speeds.

Speed	Stopping distance
10 mph	10 m
20 mph	12 m
30 mph	23 m
40 mph	36 m
50 mph	53 m
60 mph	73 m
70 mph	96 m

a Draw a graph to represent this data. Plot all the points and join them in a smooth curve.
b Estimate the stopping distance for a car travelling at 35 mph.

You should know how to ...

1 Construct functions arising from real-life problems and plot their corresponding graphs.

Check out

1 Here is a timetable for a train journey from London to Manchester:

	Arrive	Depart
London	—	0800
Milton Keynes	0830	0835
Rugby	0900	0910
Stafford	1002	1005
Manchester	1050	—

Copy the grid and draw a distance–time graph for the journey.

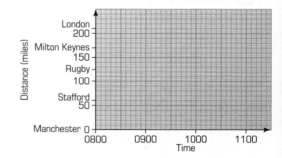

2 Interpret graphs arising from real situations.

2 Use your distance–time graph from question 1 to answer these questions.

a How far was the train from Manchester at 0930?

b How far did the train travel from London to Milton Keynes?
How long did this take?

c Find the average speed of the train from London to Milton Keynes.

1 Proportional reasoning

This unit will show you how to:

- ▶▶ Round numbers.
- ▶▶ Use prime factor decomposition of a number.
- ▶▶ Use efficient methods to add, subtract, multiply and divide fractions, interpreting division as a multiplicative inverse.
- ▶▶ Cancel common factors before calculating.
- ▶▶ Recognise when fractions and percentages are needed to compare proportions.
- ▶▶ Solve problems involving percentage changes.
- ▶▶ Use proportional reasoning to solve a problem.
- ▶▶ Compare two ratios.
- ▶▶ Interpret and use ratio in a range of contexts.
- ▶▶ Understand the effects of multiplying and dividing by numbers between 0 and 1.

- ▶▶ Use the laws of arithmetic and inverse operations.
- ▶▶ Understand the order of precedence and effect of powers.
- ▶▶ Make and justify estimates and approximations of calculations.
- ▶▶ Use known facts to derive unknown facts.
- ▶▶ Extend mental methods of calculation.
- ▶▶ Solve increasingly demanding problems and evaluate solutions.
- ▶▶ Solve problems using a range of efficient methods.
- ▶▶ Present a concise, reasoned argument.
- ▶▶ Give solutions to an appropriate degree of accuracy.

Percentages are used in everyday life.

Before you start

You should know how to ...

1 Find the HCF and LCM of two numbers.

2 Find equivalent fractions.

3 Convert between fractions, decimals and percentages.

4 Simplify a ratio.

Check in

1 The factors of 12 can be written as: $2 \times 2 \times 3$
   The factors of 40 can be written as: $2 \times 2 \times 2 \times 5$
   Find the HCF and LCM of 12 and 40.

2 Which of these pairs of fractions are equivalent?
   $\frac{1}{2}$ $\frac{24}{40}$ $\frac{2}{3}$ $\frac{15}{60}$ $\frac{6}{10}$ $\frac{12}{24}$ $\frac{3}{12}$ $\frac{24}{36}$

3 Express:
   a 20% as a fraction   **b** 0.45 as a percentage
   c $\frac{3}{20}$ as a decimal   **d** $\frac{11}{25}$ as a percentage

4 Simplify:
   a $5:15$ **b** $100:30$ **c** $36:60$ **d** $2\,km:10\,m$

Adding and subtracting fractions

This spread will show you how to:

▶▶ Use prime factor decomposition of a number.

▶▶ Use efficient methods to add and subtract fractions.

KEYWORDS

Equivalent

Denominator

Compare

Cancel

Simplify

Prime

Lowest Common Multiple

Highest Common Factor

You can use prime factors to find the highest common factor (HCF) and lowest common multiple (LCM) of two numbers:

$$18 = 2 \times 3 \times 3 \qquad 24 = 2 \times 2 \times 2 \times 3$$

$$\text{HCF} = 2 \times 3 = 6 \qquad \text{LCM} = 2 \times 2 \times 3 \times 2 \times 3 = 72$$

▶ You can add, subtract, compare and order fractions with different denominators by finding equivalent fractions with a common denominator.

You use the LCM.

example

a What is $\frac{13}{18} + \frac{17}{24}$?

b Which is larger: $2\frac{9}{56}$ or $\frac{151}{70}$?

a The LCM of 18 and 24 is 72. Find equivalent fractions with denominator 72.

$$\frac{13}{18} \xrightarrow{\times 4}{\underset{\times 4}{=}} \frac{52}{72} \qquad \frac{17}{24} \xrightarrow{\times 3}{\underset{\times 3}{=}} \frac{51}{72}$$

$$\frac{13}{18} + \frac{17}{24} = \frac{52}{72} + \frac{51}{72}$$

$$= \frac{52 + 51}{72}$$

$$= \frac{103}{72}$$

$$= 1\frac{31}{72}$$

b Using equivalent fractions,

$$2\frac{9}{56} = \frac{121}{56} \xrightarrow{\times 5}{\underset{\times 5}{=}} \frac{605}{280}$$

$$\frac{151}{70} \xrightarrow{\times 4}{\underset{\times 4}{=}} \frac{604}{280}$$

$$\frac{605}{280} > \frac{604}{280}$$

so $\quad 2\frac{9}{56} > \frac{151}{70}$

You can simplify (cancel) fractions by dividing by the HCF.

example

Simplify $\frac{105}{180}$.

The highest common factor of 105 and 180 is 15.

$$\frac{105}{180} \xrightarrow{\div 15}{\underset{\div 15}{=}} \frac{7}{12}$$

Using prime factors:

$105 = 3 \times 5 \times 7$

$180 = 2 \times 2 \times 3 \times 3 \times 5$

$\text{HCF} = 3 \times 5 = 15$

Exercise N1.1

1 Work out these answers as mixed numbers where appropriate.

a $\frac{3}{7} + \frac{5}{7}$ **b** $2\frac{3}{5} - \frac{6}{5}$

c $\frac{2}{5} + \frac{3}{10}$ **d** $\frac{2}{3} - \frac{1}{5}$

e $\frac{5}{8} + \frac{1}{2}$ **f** $\frac{7}{9} - \frac{2}{5}$

g $\frac{7}{10} + 1\frac{1}{4}$ **h** $3\frac{2}{7} - \frac{5}{8}$

2 Puzzle

Choose two fractions from the fraction box that have:

Fraction box

$\frac{2}{3}$	$\frac{1}{2}$	$\frac{1}{4}$	$\frac{2}{5}$
$\frac{3}{4}$	$\frac{1}{3}$	$\frac{3}{5}$	$\frac{5}{6}$

a a total of $\frac{5}{6}$
b a total greater than 1
c a difference of $\frac{1}{12}$
d a sum of $\frac{17}{20}$
e a total greater than $1\frac{1}{4}$ but less than $1\frac{1}{3}$.

3 Investigation

a Use four different digits to make a fraction sum with a total of 1.

$$\frac{\square}{\square} + \frac{\square}{\square} = 1$$

b Investigate different possible pairs of fractions with a sum of 1 that use four different digits.

4 Work out these answers as mixed numbers where appropriate.

a $\frac{3}{10} + \frac{3}{4}$ **b** $\frac{13}{15} - \frac{3}{5}$

c $\frac{11}{12} + \frac{7}{9}$ **d** $\frac{17}{24} + \frac{11}{18}$

e $\frac{17}{20} + \frac{-8}{15}$ **f** $1\frac{7}{12} + \frac{7}{10}$

g $2\frac{7}{18} - 1\frac{11}{12}$ **h** $\frac{23}{16} - \frac{-8}{15}$

5 For each pair of numbers, insert >, < or = in between them. Show your working out clearly for each question.

a $\frac{11}{15}$ $\frac{7}{9}$ **b** $\frac{-7}{18}$ $\frac{-13}{30}$

c $\frac{29}{12}$ $2\frac{2}{5}$ **d** $\frac{17}{24}$ $\frac{39}{56}$

e $\frac{47}{64}$ $\frac{83}{112}$ **f** $\frac{19}{51}$ $\frac{47}{136}$

6 a Rukia mixed $2\frac{5}{18}$ litres of blue paint with $1\frac{5}{15}$ litres of white paint to make a 'sky-blue' paint.
How many litres of 'sky-blue' paint did she make?

b Joshua pours $3\frac{4}{11}$ pints from a container holding $7\frac{1}{4}$ pints.
How much liquid is left in the container?

c An envelope is $6\frac{7}{8}$ inches long and $4\frac{11}{15}$ inches wide.
What is the perimeter of the envelope?

7 Puzzle

In these pyramids, the brick that sits directly above two bricks is the sum of these two bricks.
Copy and complete these pyramids:

a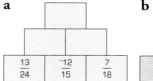

b

8 Puzzle

a The difference between two numbers is $\frac{47}{48}$.
The sum of the two numbers is $\frac{19}{48}$.
What are the two numbers?
Explain clearly the method you have used to solve the problem.

b The numbers x, $\frac{13}{18}$, y, z and $1\frac{2}{9}$ are in increasing order of size.
The difference between each successive pair of numbers is the same.
What are the values of x, y and z?
Explain clearly the method you have used to solve the problem.

Multiplying by fractions

This spread will show you how to:
- ▶▶ Use efficient methods to multiply fractions.
- ▶▶ Cancel common factors before multiplying fractions.
- ▶▶ Extend mental methods of calculation.

KEYWORDS
Integer
Numerator   Unit fraction
Improper fraction

A unit fraction has a numerator of 1: $\frac{1}{2}, \frac{1}{3}, \frac{1}{4}$

You can use unit fractions to multiply a fraction by an integer.

$$\frac{4}{7} \times 3 = (\frac{1}{7} \times 4) \times 3$$
$$= \frac{4 \times 3}{7}$$
$$= \frac{12}{7} = 1\frac{5}{7}$$

Remember:
Multiplying by $\frac{1}{7}$ is the same as dividing by 7.

You can simplify the product of a fraction and an integer ...

... at the end of the calculation

$$\frac{5}{12} \times 18 = \frac{5 \times 18}{12} = \frac{90}{12} = \frac{15}{2} = 7\frac{1}{2}$$
(÷6 arrows shown)

... during the calculation

$$\frac{5}{12} \times 18 = \frac{5 \times \cancel{18}^3}{\cancel{12}_2} = \frac{5 \times 3}{2} = \frac{15}{2} = 7\frac{1}{2}$$

You can multiply a fraction by another fraction:

$$\frac{4}{5} \times \frac{2}{3} = (4 \times \frac{1}{5}) \times (2 \times \frac{1}{3}) = \frac{4 \times 2}{5 \times 3} = \frac{8}{15}$$

multiplying by $\frac{1}{5}$ = dividing by 5   multiplying by $\frac{1}{3}$ = dividing by 3

You should always write fractions as improper fractions before multiplying.

example

Calculate:

a $\frac{18}{4} \times \frac{12}{15}$

b $1\frac{1}{4} \times 1\frac{1}{2}$

....

a $\frac{^6\cancel{18} \times ^3\cancel{12}}{_1\cancel{4} \times _5\cancel{15}}$

$= \frac{6 \times 3}{1 \times 5}$

$= \frac{18}{5}$

$= 3\frac{3}{5}$

b $1\frac{1}{4} \times 1\frac{1}{2} = \frac{5}{4} \times \frac{3}{2}$

$= \frac{5 \times 3}{4 \times 2}$

$= \frac{15}{8}$

$= 1\frac{7}{8}$

Exercise N1.2

1 Calculate these, giving your answer as a mixed number where appropriate:

   a $10 \times \frac{5}{8}$ **b** $18 \times \frac{7}{12}$ **c** $\frac{5}{7} \times 28$

   d $\frac{3}{8} \times 12$ **e** $14 \times \frac{5}{6}$ **f** $35 \times \frac{13}{28}$

   g $\frac{4}{24} \times 18$ **h** $3\frac{2}{5} \times 10$ **i** $\frac{3}{10} \times 8$

   j $\frac{9}{100} \times 4$

2 Calculate these fractions of amounts using an appropriate method:

   a $\frac{2}{5}$ of 18   **b** $\frac{7}{8}$ of 184 cm

   c $\frac{5}{13} \times £78$ **d** $\frac{11}{12} \times 20$ kg

   e $\frac{3}{7}$ of 120 cm   **f** $\frac{9}{4} \times 92$ inches

   g $2\frac{5}{9}$ of 234 cows   **h** $1\frac{4}{9}$ of £25

3 Calculate each answer in its simplest form:

   a $\frac{2}{3} \times \frac{4}{5}$ **b** $\frac{5}{7} \times \frac{14}{8}$ **c** $\frac{12}{5} \times \frac{15}{4}$

   d $\frac{6}{8} \times \frac{4}{9}$ **e** $\frac{9}{16} \times \frac{12}{3}$ **f** $\frac{8}{15} \times \frac{10}{12}$

   g $24 \times \frac{7}{18}$ **h** $\frac{18}{14} \times \frac{35}{27}$

4 Puzzle

Choose two fractions from the fraction box that have:

Fraction box			
$\frac{4}{7}$	$\frac{15}{8}$	$\frac{21}{12}$	$\frac{28}{25}$
$\frac{16}{57}$	$\frac{19}{14}$	$\frac{63}{10}$	$\frac{18}{5}$

   a a product of exactly 1
   b a product of exactly $1\frac{73}{95}$
   c a product that is less than $\frac{1}{4}$
   d the largest product
   e a product greater than $1\frac{3}{4}$ but less than 2.

5 Calculate each answer in its simplest form:

   a $1\frac{2}{3} \times 1\frac{3}{5}$ **b** $2\frac{5}{7} \times 2\frac{5}{8}$

   c $1\frac{7}{9} \times \frac{27}{32}$ **d** $1\frac{1}{15} \times 2\frac{5}{8}$

   e $\frac{28}{9} \times 1\frac{5}{7}$ **f** $2\frac{3}{16} \times 1\frac{13}{15}$

   g $1\frac{7}{11} \times \frac{44}{45}$ **h** $4\frac{1}{2} \times 1\frac{13}{27}$

   i $(2\frac{1}{4})^2$ **j** $\frac{3}{4} \times (2\frac{1}{5} - \frac{19}{45})$

6 Work out these, leaving your answers as mixed numbers where appropriate:

   a An envelope is $6\frac{3}{7}$ inches long and $4\frac{1}{5}$ inches wide.
       What is the area of the envelope?

   b The body weight of Walter the walrus is $1\frac{4}{21}$ tonnes. His tusks weigh $\frac{7}{90}$ of his total body weight.
       How heavy are Walter's tusks?

   c A jug holds $3\frac{3}{5}$ litres of orange squash. Lucy pours out $\frac{11}{12}$ of the orange squash into 6 glasses. How much orange squash is left in the jug? How much orange squash is in each glass?

7 Puzzle

Joanne spends $\frac{2}{5}$ of each weekday at work. During her work she spends $\frac{15}{22}$ of her time dealing with customers. She spends $\frac{3}{4}$ of her remaining time at work answering the phone.
What fraction of Joanne's day is spent:
   a dealing with customers
   b answering the phone?

8 Find the missing fraction in each of these calculations:

   a $\frac{4}{11} \times$ ___ $= \frac{1}{2}$

   b $\frac{5}{12} \times$ ___ $= \frac{3}{4}$

   c $\frac{7}{8} \times$ ___ $= \frac{2}{5}$

Explain your method for finding each of the missing fractions.

9 Investigation

   a Investigate what happens when you multiply:

     $\frac{1}{2} \times \frac{3}{4}$ $\frac{1}{2} \times \frac{3}{4} \times \frac{5}{6}$ $\frac{1}{2} \times \frac{3}{4} \times \frac{5}{6} \times \frac{7}{8}$ ...

   b Investigate what happens when you multiply other fractions:

     For example $\frac{1}{2} \times \frac{2}{3} \times \frac{3}{4} \times \frac{4}{5} \times$ ...

                       $\frac{1}{2} \times \frac{1}{4} \times \frac{1}{8} \times \frac{1}{16} \times$ ...

N1.3 Dividing by fractions

This spread will show you how to:
- ▶▶ Use efficient methods to divide fractions, interpreting division as a multiplicative inverse.
- ▶▶ Cancel common factors before dividing.
- ▶▶ Understand the effects of multiplying and dividing by numbers between 0 and 1.

KEYWORDS
Integer
Unit fraction
Multiplicative inverse

You can use unit fractions to divide an integer by a fraction:

$$1 \div \tfrac{1}{4} = 4$$

Think: How many quarters in 1 whole?

$$2 \div \tfrac{1}{4} = 8$$

Think: How many quarters in 2 wholes?

Multiplication and division are inverse operations:

Think: How many $\tfrac{1}{4}$s in 2?

and

Think: What is $\tfrac{1}{4}$ of 8?

Dividing by $\tfrac{1}{4}$ is the same as multiplying by 4.

Multiplying by $\tfrac{1}{4}$ is the same as dividing by 4.

▶ **You can divide an integer by any fraction.**

example

Calculate $3 \div \tfrac{3}{4}$

Think: How many $\tfrac{3}{4}$s in 3?

$$3 \div \tfrac{1}{4} = 12 \quad \text{so} \quad 3 \div \tfrac{3}{4} = 4$$

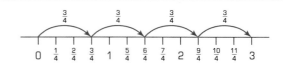

You can replace the division by the multiplicative inverse:

 and

Dividing by $\tfrac{3}{4}$ is the same as multiplying by $\tfrac{4}{3}$.

Multiplying by $\tfrac{3}{4}$ is the same as dividing by $\tfrac{4}{3}$.

example

Calculate:

a $5 \div \tfrac{4}{7}$

b $\tfrac{2}{5} \div \tfrac{3}{4}$

a $5 \div \tfrac{4}{7} = 5 \times \tfrac{7}{4}$
$= \tfrac{5 \times 7}{4}$
$= \tfrac{35}{4} = 8\tfrac{3}{4}$

b $\tfrac{2}{5} \div \tfrac{3}{4} = \tfrac{2}{5} \times \tfrac{4}{3}$
$= \tfrac{2 \times 4}{5 \times 3}$
$= \tfrac{8}{15}$

Exercise N1.3

1 Copy and complete these fraction division patterns:

a $1 \div \frac{1}{10} = 10$

$2 \div \frac{1}{10} = 20$

$3 \div \frac{1}{10} =$

$4 \div \frac{1}{10} =$

$5 \div \frac{1}{10} =$

b $1 \div \frac{1}{5} = 5$

$2 \div \frac{1}{5} = 10$

$3 \div \frac{1}{5} =$

$4 \div \frac{1}{5} =$

$5 \div \frac{1}{5} =$

c $1 \div \frac{1}{8} = 8$

$2 \div \frac{1}{8} =$

$3 \div \frac{1}{8} =$

2 Use number lines to calculate:

a $4 \div \frac{1}{3}$ **b** $2 \div \frac{2}{5}$ **c** $10 \div \frac{5}{6}$

3 Copy and complete:

a
$$5 \xrightarrow{\div \frac{1}{3}} 15 \quad \text{and} \quad 5 \xrightarrow{\times ?} 15$$
$$5 \xleftarrow{\times ?} 15 \qquad 5 \xleftarrow{\div 3} 15$$

b
$$4 \xrightarrow{\div \frac{2}{5}} 10 \quad \text{and} \quad 4 \xrightarrow{\times \frac{5}{2}} 10$$
$$4 \xleftarrow{\times ?} 10 \qquad 4 \xleftarrow{\div ?} 10$$

c
$$3 \xrightarrow{\div \frac{2}{5}} ? \quad \text{and} \quad 3 \xrightarrow{\times ?} ?$$
$$3 \xleftarrow{\times \frac{2}{5}} ? \qquad 3 \xleftarrow{\div ?} ?$$

4 Calculate, giving your answer as a mixed number where appropriate:

a $8 \div \frac{3}{4}$ **b** $12 \div \frac{5}{6}$ **c** $21 \div \frac{3}{7}$

d $15 \div \frac{2}{5}$ **e** $2 \div \frac{5}{8}$ **f** $2 \div \frac{11}{12}$

g $4 \div \frac{3}{5}$ **h** $6 \div \frac{2}{7}$ **i** $6 \div \frac{3}{10}$

5 a Lynton is a long-distance runner. He can run every kilometre in $\frac{2}{15}$ of an hour. If he runs for 2 hours, how many kilometres has he covered?

b Dave makes 4 kg of bread dough. He uses $\frac{3}{5}$ kg of bread dough to make a large loaf. How many loaves of bread can he make?

6 **Puzzle**

Michael and Sean are working out the calculation: $\frac{1}{2} \div \frac{1}{6} = ?$

Michael thinks the answer is $\frac{1}{12}$.

Sean disagrees.

Write why you think Sean disagrees with Michael's answer. Use diagrams and number lines to support your explanation.

7 Calculate, giving your answer in its simplest form:

a $\frac{1}{2} \div \frac{1}{5}$ **b** $\frac{1}{5} \div \frac{1}{4}$ **c** $\frac{2}{3} \div \frac{3}{4}$

d $\frac{3}{4} \div \frac{1}{2}$ **e** $\frac{4}{5} \div \frac{3}{7}$ **f** $\frac{7}{10} \div \frac{2}{5}$

g $\frac{4}{3} \div \frac{5}{8}$ **h** $\frac{5}{2} \div \frac{4}{9}$ **i** $1\frac{1}{4} \div \frac{3}{5}$

j $2\frac{1}{6} \div \frac{2}{3}$ **k** $6\frac{1}{2} \div \frac{7}{10}$ **l** $4\frac{2}{3} \div \frac{7}{8}$

8 **Investigation**

Write three division calculations involving fractions with an answer of $\frac{2}{3}$.

For example: $\frac{7}{12} \div \frac{7}{8} = \frac{2}{3}$

Explain how you worked out your calculations.

Use your results to solve this problem:

$\frac{4}{5} \div ? = \frac{2}{3}$

9 **Puzzle**

John has a set of operations:

He applies each operation to the number 12, e.g. $12 \times \frac{1}{2} = 6$

He says he can predict if an operation will give an answer bigger or smaller than 12 without doing any calculating.

a Without calculating, identify which operations will give an answer bigger than 12. Explain and justify your answer.

b Check by calculating the answers.

This spread will show you how to:
▶▶ Round numbers.
▶▶ Solve problems involving percentage changes.
▶▶ Use known facts to derive unknown facts.

KEYWORDS
Percentage   Increase
Fraction   Decrease
Unitary method

You should be able to use a range of methods to calculate with percentages or fractions.

▶ Finding 10% is the same as finding $\frac{1}{10}$.
▶ Multiplying by $\frac{1}{10}$ is the same as dividing by 10.

Mental methods

example

a Find 90% of £3.50

b Calculate $\frac{3}{5}$ of 23.5

a 100% of £3.50 = £3.50
      −10% of £3.50 = £0.35
      ─────────────────────
      90% of £3.50 = £3.15

b $\frac{3}{5}$ of 23.5 = $3 \times \frac{1}{5} \times 23.5$
       = $3 \times (23.5 \div 5)$
       = 3×4.7
       = 14.1

Equivalent calculations

The unitary method

example

a Find 2.3% of 48.6 kg

b Find 117% of £45

a 2.3% of 48.6 kg = $\frac{2.3}{100} \times 48.6$ kg
       = $(2.3 \div 100) \times 48.6$ kg
       = 0.023×48.6 kg
       = 1.1178 kg
       = 1.1 kg (1 dp)

b 117% of £45 = (1% of £45) × 117
       = ($\frac{1}{100}$ of £45) × 117
       = £0.45 × 117
       = £52.65

▶ You can calculate a percentage change in a single calculation.

If something is decreased by 9% then it is 91% (100% − 9%) of its original size.

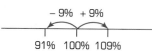
− 9% + 9%
91% 100% 109%

If something is increased by 9% then it is 109% (100% + 9%) of its original size.

example

When frozen, 275 cm³ of water expands in volume by 9%.
What is the volume of the ice?

Volume of ice = (100 + 9)% of the original volume of water
= 109% of 275 cm³
= 1.09 × 275 cm³
= 299.75 cm³ = 300 cm³ (nearest cm³)

Exercise N1.4

1 Use an appropriate method to calculate:
  a 10% of £380   **b** $\frac{2}{5}$ of 230 m
  c 65% of 48 kg   **d** $1\frac{1}{5}$ of 65 DVDs
  e 12% of 160 cm   **f** $\frac{7}{20} \times 260$ glasses
  g 99% of 450 kg   **h** 125% of 3.8 km

2 **a** Ian used $\frac{1}{5}$ of a 480 g bag of rice to make a risotto. Clyde used $\frac{2}{3}$ of the rice that remained to make a rice pudding. How many grams of rice were left in the bag?

   b Avril has £4500 to spend on a holiday. She uses 34% of the money to pay for the airfares. She spends 54% of the remaining money on luxury accommodation. How much spending money does she have left?

3 Calculate, giving your answer as a fraction or as a decimal to 1 dp where appropriate:
  a 28% of 72 kg   **b** 123% of 35 km
  c $\frac{3}{8}$ of 7.8 cm   **d** 2.5% of 64 mm
  e $2\frac{5}{7} \times 85$ litres   **f** 7.6% of 40 mm
  g 17.5% of £81   **h** $\frac{9}{25}$ of 34 hectares
  i 171% of 7.5 m

4 **Puzzle**
  a Lucy can buy a washing machine for one cash payment of £329, or pay using the '36-plan' option. With the '36-plan' she pays a deposit of 36% and then six equal monthly payments of £36. How much extra does she pay by the '36-plan' option?

  b Which is the better buy: a 50 g packet of crisps at 42p, or a packet of crisps with 50 g + 30% extra at 54p?

  c In a survey of 140 pupils, $\frac{2}{7}$ liked football best, $\frac{1}{5}$ liked cricket, and $\frac{3}{10}$ liked athletics. The rest liked cycling. How many liked cycling?

5 Calculate, giving your answers to 2 dp as appropriate:
  a A pair of designer jeans cost £62 in GIP Clothing shop. In a sale, all prices in the shop are reduced by 15%. What is the new price of the jeans?

  b A digital TV costs £345 plus a charge of 17.5% VAT. What is the total cost of the digital TV?

  c A joint of meat weighs 2.5 kg when frozen. It loses 7% of its weight when it is thawed. What is the weight of the meat when it is thawed?

  d If you invest more than £2000 in a Bigsaver account you will receive 3.9% interest a year. Roger invests £3400 in the Bigsaver account. How much interest will he receive after 1 year?

  e A metal bar is 4.6 m long. When heated it expands in length by 1.2%. What is the length of the bar after it has expanded?

  f When using his weighing scales, Ernie knows that his instruments have an error of up to 1.5%. If he weighs 3 kg of carrots using his scales, what is the:
    i Greatest possible weight the carrots could be
    ii Least possible weight the carrots could be?

6 **Puzzle**
Bill Eatalot goes to a restaurant. At the end of the meal the waiter adds VAT at 17.5% and then a service charge of 8%. Bill tells the waiter that he should have added the service charge first and then added the VAT.
Whose method was cheaper? Explain and justify your answer.

This spread will show you how to:

▶▶ Recognise when fractions and percentages are needed to compare proportions.

▶▶ Solve problems involving percentage changes.

A fraction is the result of a division.

$$\frac{3}{8} = 3 \div 8 = 0.375 \qquad \frac{13}{7} = 13 \div 7 = 1\frac{6}{7}$$

13 ÷ 7 = 1·571428571 …
which is a recurring decimal.

▶ You use fractions and percentages to compare proportions.

example

a What proportion of the square is shaded?

b The election results in Treehorserace were:

Labour   33.3%
Liberal Democrat   823 votes
Conservative   41 in every 124 votes
Others   11 votes

There were 2480 votes cast in the election. Who won the election?

a There are 8 equal-sized triangles.
The fraction shaded = $\frac{5}{8}$
The percentage = $(5 \div 8) \times 100\%$
                = 62.5%

b The percentage of votes for each party is:
L   33.3%
LD $(823 \div 2480) \times 100\% = 33.2\%$ (1 dp)
C $(41 \div 124) \times 100\% = 33.1\%$ (1 dp)
O $\left(\frac{11}{2480}\right) \times 100\% = 0.4\%$ (1 dp)
Labour won the election with the highest proportion of votes.

When a quantity changes you can calculate the percentage change.
You compare the change with the original amount.

▶ % Change = $\dfrac{\text{change}}{\text{original amount}} \times 100\%$

The change can be an increase or a reduction.

example

a A fridge is reduced from £440 to £374. What is the percentage reduction?

b A TV increases from £1320 to £1386. What is the percentage increase?

a Reduction   = £440 − £374 = £66

% Reduction = $\frac{66}{440} \times 100\%$
            = 0.15 × 100% = 15%

The price has been reduced by 15%.

b Increase   = £1386 − £1320 = £66

% Increase = $\frac{66}{1320} \times 100\%$
            = 0.05 × 100% = 5%

The price has been increased by 5%.

Exercise N1.5

1 What fraction of:
   a 120 is 35   **b** 4 days is 8 hours   **c** 5 kg is 2400 g   **d** £110 is £12
   e 240 paper clips is 7 paper clips?
   i Write your answers as fractions expressed in their simplest form.
   ii Write your answers as percentages (to 1 dp where appropriate).

2 This is the nutritional information for a
   750 g packet of breakfast cereal.
   a Calculate the proportion of protein,
      carbohydrate, fat and fibre in the packet
      (give your answers to 1 dp where
      appropriate).
   b How much fibre would you expect to
      find in a 30 g serving?

Nutritional information	
Composition	750 g provides
Energy	2482 kcal
Protein	76.5g
Carbohydrate	503.3g
Fat	18g
Fibre	105.8g

3 Here are the heights of all the students in Year 9 from two schools:

South Park High School

Height (cm)	Less than 140	$140 \leqslant h < 150$	$150 \leqslant h < 160$	$160 \leqslant h < 170$	$170 \leqslant h < 180$	$180 \leqslant h < 190$
Frequency	3	7	40	15	6	4

Greendale Sports College

Height (cm)	Less than 140	$140 \leqslant h < 150$	$150 \leqslant h < 160$	$160 \leqslant h < 170$	$170 \leqslant h < 180$	$180 \leqslant h < 190$
Frequency	7	16	59	21	7	5

   a Which school has the highest proportion of students between
      150 cm and 160 cm in height?
   b Work out the percentage of students in each height range for both schools.
   c Which school has the taller Year 9 students? Explain and justify your answer.

4 Calculate, giving your answers to 1 dp where appropriate:
   a The cost price of a car radio is £245. Its selling price is £196.
      What is the percentage reduction?
   b A piece of elastic is 64 cm long. It is stretched to a length
      of 73.6 cm. Find the percentage change in its length.
   c At Bigschool Technology College, the number of boys who
      obtained a level 5 or better in their KS3 tests increased
      from 124 (in 2002) to 146 (in 2003). By what percentage
      has the number of boys obtaining a level 5 or better increased?
   d DOTCOM2 shares cost 24p each in 2001. Three years later they rose in price to 32p.
      By what percentage has the cost of the shares risen?
   e A pair of RUNFAST trainers are reduced in price from £49 to £28.
      By what percentage has the price of the trainers been reduced?

Percentage change

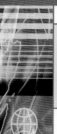

This spread will show you how to:
▶▶ Solve problems involving percentage changes.
▶▶ Use proportional reasoning to solve a problem.
▶▶ Use the laws of arithmetic and inverse operations.

KEYWORDS
Unitary method   Inverse
Multiplicative inverse

A coat is on sale for £45.50.

It has been reduced in price by 35% from its original price.

The coat is now (100% − 35%) = 65% of its original price.

Here are some methods of finding the original price:

35% off

Sale Price
£45.50

The unitary method

Find the value of 1% of the original price.

$\times \frac{1}{65}$ ⎧ £45.50 is 65% of the original price
         ⎨ £0.70   is 1% of the original price $\times \frac{1}{65}$
$\times 100$ ⎩ £70.00 is 100% of the original price $\times 100$

Multiplying by $\frac{1}{65}$ is the same as dividing by 65.

Using inverse relationships

You know that ...

| Reduced price = 65% of original price |
| Reduced price = $\frac{65}{100}$ × original price |
| Reduced price = 0.65 × original price |

You can say that ...

| Original price = $\frac{100}{65}$ × reduced price |
| Original price = 1.54 × reduced price |
| Original price = 154% of reduced price |

× 0.65

original price   reduced price

÷ 0.65

Original price = reduced price ÷ 0.65
            = £45.50 ÷ 0.65
            = £70.00

$\times \frac{65}{100}$

original price   reduced price

$\times \frac{100}{65}$

$\frac{100}{65}$ is the multiplicative inverse of $\frac{65}{100}$.

Original price = reduced price × $\frac{100}{65}$
            = $\frac{£45.50 \times 100}{65}$
            = £70.00

Notice that when you divide by 0.65 the answer is larger!

---

example

A shirt is reduced in price by 20%.
The sale price is £20.
Calculate the original price.

Sale price = 80% of original price
        = $\frac{80}{100}$ × original price

Original price = $\frac{100}{80}$ × new price
            = $\frac{5}{4}$ × £20
            = £25

Exercise N1.6

1 Calculate:
   a 15% of 64 kg
   b 8% of £1750
   c 5% of 27 m
   d 41% of 23.5 litres
   e $\frac{4}{7}$ of 336 gallons
   f 1% of 12.6 hectares.

2 Investigation
   a Mr Brown increases all the prices in his shop by 10%.
   Copy and complete this price list

Item	Original price	New price
Bread	80p	88p
Milk (1 litre)	60p	
Bran Flakes	£1.80	
Eggs (doz)	£1.30	

   b The new price of a 2 kg packet of frozen peas is £2.75.
   What was the original price?
   c Investigate how to find the original price of different items after a 10% increase.

3 In a sale, all prices are reduced by 20%. The sale price is 80% of the original price.

   a Use this information to copy and complete this conversion table:

Original price	Sale price
£10.00	
£16.99	
	£20.00
	£6.00

   b Explain your method for changing the sale price into the original price.

4 a A pair of trainers are on sale for £45 which is 75% of the original price. What was the original price of the trainers?
   b About 41 600 people visited the mini-golf course in 2003. This was an increase of 30% on 1998. How many visitors to the mini-golf course were there in 1998?
   c Emma sells her house for £92 400, making a profit of 65% on the amount she paid for it. What was the price she paid for the house?
   d A banner on a packet of biscuits says that it is 43% bigger. It contains 40 biscuits. How many biscuits are there in the normal packet?
   e Gareth follows a diet for 3 months. At the end of the 3 months he weighs 88 kg, which is 8% less than his weight before the diet. What did Gareth weigh before his diet?

5 Puzzle
   a Ivor bought a game-box video game console and 4 games in a sale and saved £48. The label said that it was a 15% reduction. What was the original price of the game-box and 4 games?
   b Karen's salary rises by 2.5%. This means that she earns an extra £450 each year. What is Karen's new annual salary after the pay rise?

6 Puzzle
   After a TV advertising campaign costing £50 000, Cars-R-Us car dealers found that their profits had risen by 12% to £488 000. From a financial point of view, was the advertising worthwhile? Justify your answer.

This spread will show you how to:
- ▶▶ Compare two ratios.
- ▶▶ Interpret and use ratio in a range of contexts.

KEYWORDS

Ratio   Compare
Scale factor

A ratio compares the size of parts with each other.
You usually use whole numbers in a ratio.

example

A fruit drink is made in the ratio $2\frac{3}{5}$ parts pineapple to 4 parts grapefruit.
Express the ratio in whole number form.

Pineapple : Grapefruit   = $\times 5 \left(\begin{array}{c} 2\frac{3}{5} : 4 \end{array} \right) \times 5$

=   13 : 20

▶ **You can write a ratio as a scale factor.**

The scale factor tells you how many times bigger one number is compared with another.

You can calculate the size of a scale factor using division.

Bob weighs 1.5 kg   Flipper weighs 4 kg

Bob's weight : Flipper's weight
= 1.5 kg : 4 kg
= 3 : 8

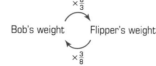

Bob's weight   Flipper's weight

Bob : Flipper	Scale factor $= \dfrac{\text{Bob's weight}}{\text{Flipper's weight}}$
	$= \frac{1.5}{4} = \frac{3}{8} = 0.375$

Flipper : Bob	Scale factor $= \dfrac{\text{Flipper's weight}}{\text{Bob's weight}}$
	$= \frac{4}{1.5} = \frac{8}{3} = 2.667$

Bob's weight $= \frac{3}{8} \times$ Flipper's weight
Bob's weight $= 0.375 \times$ Flipper's weight
Bob's weight is 37.5% of Flipper's weight
For every 1 kg of Flipper there is 0.375 kg of Bob

Flipper's weight $= \frac{8}{3} \times$ Bob's weight
Flipper's weight $= 2.667 \times$ Bob's weight
Flipper's weight is 266.7% of Bob's weight
For every 1 kg of Bob there is 2.667 kg of Flipper

Unitary ratios help you compare amounts more easily.

A unitary ratio is written in the form $1 : m$.

example

The ratio of a map is given as 5 cm : 400 m. Express this ratio in the form $1 : m$.

Ratio =   5 cm : 400 m

=   5 cm : 40 000 cm

=   5 : 40 000

$\div 5 \left(\begin{array}{c} \\ 1 : 8000 \end{array} \right) \div 5$

Real-life distance = 8000 × map distance.

Every 1 cm on the map represents 8000 cm in real life.

Exercise N1.7

1 Simplify these ratios.
  a $12 : 50$　　　　**b** $15 : 18 : 24$
  c $3 \text{ m} : 50 \text{ cm}$　　**d** $800 \text{ g} : 3 \text{ kg}$
  e $150 \text{ cm} : 500 \text{ mm}$

2 **a** In a triangle, the ratio of the angles is $2 : 5 : 11$. Calculate the angles.
  b Axel and Morrissey shared some milk in the ratio $2 : 5$. Axel received 300 ml of milk.
   How much milk was there altogether?
  c To make orange paint, you mix 7 litres of yellow with 3 litres of red paint. How many litres of each colour do you need to make 120 litres of orange paint?

3 Express these as ratios in their simplest form.
  a $0.5 : 3$　　　　**b** $2\frac{1}{2} : 4$
  c $5.2 \text{ kg} : 6.5 \text{ kg}$　**d** $3\frac{1}{4} \text{ m} : 2 \text{ m}$
  e $4\frac{2}{5} \text{ kg} : 3 \text{ kg}$

4 Express these ratios in the form $1 : m$ or $m : 1$.
  a $7 : 56$　　　　**b** $15 : 3$
  c $25 \text{ cm} : 12 \text{ cm}$　**d** $4 \text{ cm} : 50 \text{ m}$
  e $3.5 \text{ cm} : 84 \text{ cm}$

5 **a** Granny's biscuits are made from flour and butter in the ratio $7 : 4$.
   How much butter will be needed to make 495 g of biscuits?
  b A type of plaster is made in the ratio $1\frac{3}{4}$ parts cement to 3 parts sand. How much cement needs to be mixed with 24 kg of sand to make the plaster?
  c An alloy is made from iron, copper and aluminium in the ratio $5 : 3 : 4$.
   How much iron is needed to mix with 108 kg of aluminium?

6 One day, Mike did a survey of the birds on his walk round Shibden Hall. He saw 6 finches, 12 sparrows, 15 crows and 27 pigeons.
  a Write the ratio of finches : sparrows : crows : pigeons in its simplest form.
  b What percentage of the birds seen by Mike were pigeons?
  c A week later, Mike did the same survey. This time the number of finches had decreased by a $\frac{1}{3}$, the number of sparrows had increased by $\frac{5}{6}$, the number of crows had increased by 20% and the number of pigeons had decreased by $\frac{2}{9}$. Write the new ratio of finches : sparrows : crows : pigeons in its simplest form.

7 **Puzzle**
On 24th October 1970 Peter is 120 cm tall. Ten years later his height has increased by 45%.
What is the ratio of Peter's height in 1970 compared with his height in 1980?

8 **a** 3 parts of blue paint are mixed with 4 parts of yellow paint to make green paint.
   What is the maximum amount of green paint that can be made from 60 ml of blue and 72 ml of yellow paint?
  b At a circus, the cost of an adult ticket is $1\frac{2}{3}$ the cost of a child's ticket.
   If the total cost of one adult and one child ticket is £6.40, how much does an adult ticket cost?

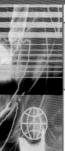

This spread will show you how to:

▶▶ Use proportional reasoning to solve a problem, choosing the correct numbers to take as a whole.

KEYWORDS

Ratio   Proportion
Scale factor   Proportional to

The table shows some pounds into euros conversions.

Pounds	Euros
6	10.5
36	63
40	70
21	36.75

The ratio of pounds to euros is always the same:

pounds : euros $= 40 : 70$
$\qquad\qquad\quad = \ \ 4 : 7$
pounds : euros $= 36 : 63$
$\qquad\qquad\quad = \ \ 4 : 7$

The number of pounds is proportional to the number of euros.

The ratio can be expressed as a **scale factor**:

$\frac{\text{euros}}{\text{pounds}} = \frac{70}{40} = 1.75$

This scale factor will change pounds into euros.

▶ Sets of numbers are proportional when the ratio of corresponding numbers is the same.

The numbers increase in the same proportion.

You can use proportionality to solve problems.

A recipe uses 35 g of sugar to make 100 g of jam.
How much sugar do you need to make 454 g of jam?

Unitary method

Find the amount of sugar needed to make 1 gram of jam:

$\times \frac{1}{100}$ $\begin{cases} 100\text{ g of jam} & \text{needs} & 35\text{ g of sugar} \\ 1\text{ g of jam} & \text{needs} & 0.35\text{ g of sugar} \\ 454\text{ g jam} & \text{needs} & \textbf{158.9 g of sugar} \end{cases}$ $\times \frac{1}{100}$
$\times 454$ $\times 454$

Scaling method

Find the proportion by which each number has changed:

Proportion $= \frac{454\text{ g}}{100\text{ g}} = \frac{454}{100} = 4.54$

$\times \frac{454}{100}$ $\begin{cases} 100\text{ g of jam} & \text{needs} & 35\text{ g of sugar} \\ 454\text{ g jam} & \text{needs} & \textbf{158.9 g of sugar} \end{cases}$ $\times \frac{454}{100}$

Scale factor method

Find the scale factor to change the number of grams of jam into the number of grams of sugar:

Scale factor
$= \frac{\text{grams of sugar}}{\text{grams of jam}}$
$= \frac{35}{100} = 0.35$

$\times 0.35$
grams of jam grams of sugar
$\div 0.35$

454 g of jam needs 454 g × 0.35 = **158.9 g of sugar.**

Exercise N1.8

1 Work out each of these using a mental or written method:
  a 4 bars of chocolate cost 80p. How much would 3 bars of chocolate cost?
  b 5 miles is approximately equal to 8 kilometres. Roughly how many kilometres are equal to 35 miles? How many miles are equal to 112 km?
  c 7 pizzas cost £20.65.
  What is the cost of 14 pizzas?
  What is the cost of 5 pizzas?

2 A 5 kg bag of apples is about the same as an 11 pound bag of apples.

  a Use this information to copy and complete this conversion table:

Kilograms	Pounds
30	
1	
	187
	14

  b What is the scale factor for changing pounds into kilograms?

3 Solve these, clearly showing your method.
  a A recipe for jam uses 45 g of fruit for every 100 g of jam. Jack wants to make five 454 g jars of jam.
  How much fruit does he need?
  b A recipe for fruit squash requires 840 ml of water for 7 people.
  How much water would you need to make fruit squash for 11 people?
  c A pair of trainers are on sale for £33.60 which is 80% of the original price. What was the original price of the trainers?

4 The cost of a take-away meal for two is £7.60. The pie chart shows how the total cost of the meal is made up.

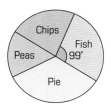

The cost of a fish is represented by a sector angle of 99° on the pie chart. What is the cost of a fish?

5 Puzzle
  a If £1 = 1.64 euros, how many euros would you get for £3.50? How many euros would you get for 85p?
  b If £1 = 10.4 rand, how many pounds would you get for 52 rand?
  How many pounds would you get for 4 rand?

6 a A train is travelling at a constant speed. After 3 hours it has travelled 255 km. How far had it travelled after 2 hours? How far had it travelled after 48 minutes?
  b A car is travelling at an average speed of 45 mph. How far does it travel in 4 hours? How far does it travel in $2\frac{1}{3}$ hours? How long will it take the car to travel 144 miles?

7 Puzzle
Here are some currency rates for different countries:

	UK	France	India
UK		£1 = 1.64 euros	£1 = 71.2 rupees
France	1 euro =		1 euro =
India	1 rupee =	1 rupee =	

Work out the missing entries.

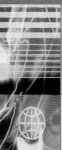

This spread will show you how to:

▶▶ Understand the order of precedence and effect of powers.

▶▶ Make estimates and approximations of calculations.

KEYWORDS

Order of operations

Square root · Power

Cube number · Index

Cube root · Brackets

You need to use the correct order of operations to work out a calculation.

example

Calculate $\dfrac{(\frac{3}{5}+3)^3}{\sqrt{(^-1.5+5\times3.5)}} + \dfrac{3.6\times0.5}{(\frac{3}{4})^2}$

Order of operations

Brackets
Work out the contents of any brackets first.

$$\frac{(\frac{3}{5}+3)^3}{\sqrt{(^-1.5+5\times3.5)}} + \frac{3.6\times0.5}{(\frac{3}{4})^2}$$

Use the order of operations inside the bracket:
$^-1.5+5\times3.5$
$=\ ^-1.5+17.5=16$

Powers or indices
Work out any powers, square roots or cube roots next.

$$=\frac{(3\frac{3}{5})^3}{\sqrt{16}} + \frac{3.6\times0.5}{(\frac{3}{4})^2}$$

$(3\frac{3}{5})^3=(3.6)^3=46.656$
$\sqrt{16}=4$
$(\frac{3}{4})^2=\frac{9}{16}=0.5625$

Multiplication and division
Work out multiplications and divisions.

$$=\frac{46.656}{4} + \frac{3.6\times0.5}{0.5625}$$

$46.656\div4=11.664$
$(3.6\times0.5)\div0.5625$
$=1.8\div0.5625$
$=3.2$

Addition and subtraction
Finally, work out any additions and subtractions.

$= 11.664 + 3.2$

$= 14.864$
$= 14.9$ (1 dp)

You need to interpret brackets and powers very carefully.

The power relates to everything inside a bracket:

7^2	$=$	7×7	$=$	49
$^-7^2$	$=$	$^-(7\times7)$	$=$	$^-49$
$(^-7)^2$	$=$	$^-7\times^-7$	$=$	49
$\frac{3^3}{4}$	$=$	$\frac{3\times3\times3}{4}$	$=$	$\frac{27}{4}$
$(\frac{3}{4})^3$	$=$	$\frac{3}{4}\times\frac{3}{4}\times\frac{3}{4}$	$=$	$\frac{27}{64}$

Adjacent brackets should be multiplied:

$(16-12)(3.4-10)=4\times^-6.6=\ ^-26.4$

On a calculator you input:

Powers using $\boxed{x^y}$

Fractions using $\boxed{a^{b\!\!/_{\!c}}}$

Negatives using $\boxed{+/-}$

Brackets using $\boxed{(}$ then $\boxed{)}$

Square roots using $\boxed{\sqrt{\ }}$

Cube roots using $\boxed{^x\sqrt{\ }}$ or $\boxed{^3\sqrt{\ }}$

Exercise N1.9

1 Calculate these, using a mental, written or calculator method, giving your answer to 2 dp where appropriate:

a $15 + 8 \times 19$ **b** $(3^2 + 5^2)^2$

c $80 \div 10 \div 2$

d $4.2 - (2.6 + 4.2) + (7 + 2.6)$

e $\frac{17 - 3}{15 - 4}$ **f** $\frac{3.7 + 2.8}{2.6}$

g $\frac{35}{8 \times 19}$ **h** $160 \div \{24 - (5^2 - 9)\}$

2 **Investigation**
Peter says that 'Adding two numbers and then squaring them is always the same as squaring each number and then adding'. Kyle says that 'Adding two numbers and then squaring them is never the same as squaring each number and then adding'. Who do you think is correct?
Explain and justify your answer.

3 Solve each of these calculations. You will need to decide whether to use a mental, written or calculator method. Where appropriate, give your answer to 2 dp.

a $^-6^2 + 11$

b $(11 - 17)(13 + 4)$

c $\left(\frac{4}{3}\right)^2$

d $\frac{(16 - 12)^2(^-6 + ^-3)^2}{72}$

e $\frac{(8 \times 5)^2}{\sqrt{(2 \times 8)}} + 14$

f $\frac{(^-7)^2 - 31}{(3 \times 3)^2}$

g $\frac{(17 - 6)^2(17 - 2)}{(3^2 - 4)}$

h $^-(302 \times 4 + 256) + 4 \times 299 - (5 - 304)$

i $15 \div [3 - 4(1 - 4)]$

j $\sqrt{(21^2 - 14^2)}$

k $\frac{(17 - 8)^2(18 - 15)^2}{3(14 - 8)^3}$

4 **a** Write down an approximate answer for each of these questions, clearly showing your method of approximation.

i $\frac{44.63 \times 72.8}{2.1 \times (5.4 - 2.9)}$

ii $3.7 + (5.5 - (12.3 \times 6.4))$

iii $\{(2.3)^2 + (6.2 - 1.38)\}^2$

iv $\frac{4 \times \sqrt{(3.7^2 + 3^2)}}{1.7}$

b Use your calculator to work out the exact answer, where appropriate giving your answer to 2 dp.

5 **Puzzle**
Copy these calculations, inserting the correct operations and brackets to make them correct.

a $\frac{1}{2}$ __ 4 __ 3 __ 8 = 28

b 20 __ 4^2 __ 6 __ = 2

c $\sqrt{[2^2}$ __ 3 __ 11 __ 5 __ 8] = 8

6 Kyle uses this formula to work out the area of a special shape:

$$\text{Area} = \tfrac{1}{2}ab + \frac{(a^2 + b^2)}{4}$$

a For his first shape, Kyle uses $a = 3.5$ and $b = 5$ and substitutes them into his formula. Work out the value of:

$$\tfrac{1}{2} \times 3.5 \times 5 + \frac{(3.5^2 + 5^2)}{4}$$

b Work out the area of a shape when $a = 2.4$ and $b = 1.3$.

You should know how to ...

1 Add, subtract, multiply and divide fractions.

2 Use proportional reasoning to solve a problem, choosing the correct numbers to take as a whole.

3 Make and justify estimates and approximations.

4 Solve problems using a range of efficient methods.

5 Present a concise, reasoned argument.

Check out

1 a Add $\frac{3}{8}$ and $\frac{7}{12}$.

  b i Calculate the value of $\frac{3}{4} \times \frac{2}{5}$.

    ii Show your answer on a copy of this number line.

  c How many eighths are there in $2\frac{3}{4}$?

  d Work out $2\frac{3}{4} \div \frac{5}{8}$.

2 A recipe for mushroom pie contains the following ingredients in every 225 g serving.

Mushrooms	153 g
Onions	40 g
Pastry	30 g
Herbs	2 g

  a How many grams of mushrooms would there be in 350 g of mushroom pie?

  b What percentage of the pie is made from onions?

3 a Explain which of the following will give answers larger than 10 and which will produce answers smaller that 10 without doing any calculations.

    i $10 \times \frac{2}{5}$ **ii** $10 \div \frac{2}{5}$ **iii** $10 \times \frac{7}{5}$ **iv** $10 \div \frac{7}{5}$

  b Find approximate answers to these questions.

  c Calculate each of the answers.

4 Check each of your answers to question **2** by:

  a Estimating or approximating the answer to each question.

  b Using a different method to solve the problem.

5 John has just worked out 85% of £340. His answer is £400.

  a Without doing any calculations, explain why John's answer must be wrong.

  b Work out the correct answer.

  c Write down what mistake you think John has made.

This unit will show you how to:

▶▶ Use rounding to make estimates.

▶▶ Distinguish between the different roles played by letter symbols in equations, identities, formulae and functions.

▶▶ Use index notation for integer powers and simple instances of the index laws.

▶▶ Construct and solve linear equations with integer coefficients.

▶▶ Use systematic trial and improvement methods to find approximate solutions.

▶▶ Solve problems involving direct proportion using algebraic methods, relating algebraic solutions to graphical representations.

▶▶ Solve increasingly demanding problems and evaluate solutions.

▶▶ Represent problems and interpret solutions in algebraic and graphical forms.

▶▶ Move from one form to another to gain a better perspective on the problem.

▶▶ Use trial and improvement where a more efficient method is not obvious.

500 kilos to port, NOW!

You also have to keep an equation in balance – the expressions on each side must be equal.

Before you start

You should know how to ...

1 Simplify algebraic expressions.

2 Solve simple equations.

3 Substitute values into formulae.

4 Expand brackets.
$3(x - 2) = 3 \times x - 3 \times 2 = 3x - 6$

Check in

1 Simplify:
   a $3x + 2y - 5x + 2x - 3y$
   b $3p - 2r + 7 - 2p + r$

2 Solve:
   a $3p + 4 = 19$ **b** $7n - 12 = 2$

3 Copy and complete the table for given values of x when $y = 2x - 1$.

x	$^-2$	$^-1$	0	1	2
y					

4 Expand and simplify:
   a $2(t - 3y)$ **b** $12 - 3(x - 4)$

Using letter symbols

This spread will show you how to:
- ▶▶ Distinguish the different roles played by letter symbols.
- ▶▶ Use index notation for integer powers.
- ▶▶ Simplify or transform algebraic expressions.

KEYWORDS

Equation   Formula
Identity (≡)   Function
Expression   Simplify
Equivalent

You can use letter symbols to stand for unknown values:

In an equation	$3x - 7 = 5$	... x has a particular value.
In a function	$y = 7 - 5x$	... you can find a unique value of y for any value of x.
In a formula	$V = IR$	... V, I and R are variable quantities. When you know two of the values you can find the third.
In an identity	$4(x + 3) \equiv 4x + 12$	... the expressions on either side of the identity always have the same value.

▶ ≡ is the identity sign that shows two expressions are equivalent.

example

Here are some algebra cards:

$n + n + n$ $3 \times n \times n$

$3n + n$ $3n$

$4(4 - n)$

a Find two equivalent cards.
b Which card is equivalent to $3n^2$?
c Which two cards add to give a total of 16?
d Write a card that is equivalent to $2n + 3(n - 2)$.

....

First simplify each card:
$n + n + n = 3n$
$3 \times n \times n = 3n^2$
$3n + n = 4n$
$4(4 - n) = 4 \times 4 - 4 \times n = 16 - 4n$

a $n + n + n \equiv 3n$
b $3 \times n \times n \equiv 3n^2$
c $4n + (16 - 4n) = 16$ so the two cards are $3n + n$ and $4(4 - n)$.
d $2n + 3(n - 2) = 2n + 3n - 6 = 5n - 6$
    so $5n - 6 \equiv 2n + 3(n - 2)$

Exercise A3.1

1 Here are some expression cards.

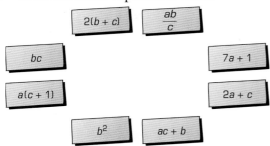

  a Work out the value of each card when $a = 5$, $b = 6$ and $c = 2$.

  b Copy the diagram and join pairs of expressions that have the same value when $a = 5$, $b = 6$ and $c = 2$.

2 Copy this diagram.
Join pairs of expressions that will always have the same value when $x = y = z$.

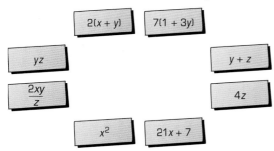

3 Simplify each of these expressions. The first is done for you.

  a $3(n+4) - 2(1+2n)$
    $= 3n + 12 - 2 - 4n$
    $= {}^-n + 10$

  b $2n + 5 + 7n$

  c $3n - 6 + 5 - n$

  d $5(2n-3) + 8n + 1$

  e $n(n+2) - 3n$

  f $5n - 2(2n+1)$

  g $\dfrac{6n^2}{2}$

  h $8n + \frac{1}{2}(12 - 10n)$

  i $2(3n+1) + 3(4n-1)$

  j $5(2n-7) - 4(n-3)$

4 Write an expression for the perimeter of each of these shapes. Find a value for x.

  a

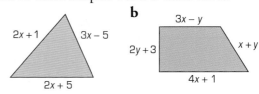

    Perimeter = 15 cm $y = 5$ cm, Perimeter = 94 cm

  c

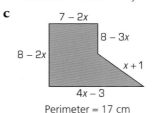

    Perimeter = 17 cm

5 **a** Find the length of each missing side.

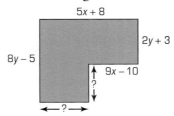

  b What is the perimeter of the shape?

6 Simplify each of these expressions.

  a $a^3 \times a^2$

  b $a^7 \div a^3$

  c $3a \times 2a^2$

  d $\dfrac{6a^3}{2}$

  e $a^2 \times 3a^2 \times 2a$

  f $\dfrac{8a^4}{a^2}$

  g $4(3a^3 - a^2)$

  h $\dfrac{4a^7}{2a^3}$

> Hint: write the expressions in full and combine or cancel fractions.

7 Work out the value of each expression in question 6 when $a = 10$.

Balancing linear equations

This spread will show you how to:
▶▶ Construct and solve linear equations.

You can solve a linear equation using the balance method.
You change both sides in the same way.

$$18 - 2x = 3x - 2$$

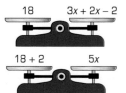

In this equation there are more xs on the right.

Add $2x$ to both sides:

	$18 - 2x + 2x$	$= 3x + 2x - 2$
	18	$= 5x - 2$

Add 2 to both sides:

	$18 + 2$	$= 5x - 2 + 2$
	20	$= 5x$

Divide both sides by 5:

	$20 \div 5$	$= 5x \div 5$
	4	$= x$

Check the solution: substitute $x = 4$ into each expression.

$$18 - 2x = 18 - 2 \times 4 = 18 - 8 = 10$$
$$3x - 2 = 3 \times 4 - 2 = 12 - 2 = 10$$

The answers are the same, so $x = 4$ is correct.

You often need to form equations to solve problems.

You can see if you have made a mistake by checking.

example

The lengths of this rectangle are in cm:

a Construct equations to find the value of x and y.
b Find the area and perimeter of the rectangle.

$3x + 10$

$20 - 3y$ ▭ $2y + 5$

$5x - 2$

a You know that opposite sides are equal.

So	$5x - 2$	$= 3x + 10$	and	$20 - 3y$	$= 2y + 5$
$(-3x)$	$2x - 2$	$= 10$	$(+3y)$	20	$= 5y + 5$
$(+2)$	$2x$	$= 12$	(-5)	15	$= 5y$
$(\div 2)$	x	$= 6$	$(\div 5)$	3	$= y$

Check: $\quad 5x - 2 = 5 \times 6 - 2 = 28 \qquad$ Check: $\quad 20 - 3y = 20 - 3 \times 3 = 11$
$\qquad\quad 3x + 10 = 3 \times 6 + 10 = 28 \qquad\qquad\qquad 2y + 5 = 2 \times 3 + 5 = 11$

So $x = 6$ cm and $y = 3$ cm

b The lengths of the sides are: 11 cm and 28 cm.
The area is 11 cm × 28 cm = 308 cm^2.
The perimeter is 11 cm + 11 cm + 28 cm + 28 cm = 78 cm.

Exercise A3.2

1 Work out the value of x in each of these equations
(use the balance method or another appropriate method).

   a $x + 5 = 9$ **b** $3x = 15$ **c** $16 = 8 + x$ **d** $3x - 5 = 28$

   e $5 + 2x = 13$ **f** $6x - 7 = 53$ **g** $18 - 2x = 4$ **h** $\frac{x}{4} + 1 = 5$

   i $2x + 1 = 3x - 2$ **j** $5 + 6x = 2x + 21$ **k** $9x + 3 = 45 - 12x$ **l** $\frac{3x}{2} - 7 = 11$

2 Use the balance method to solve these equations.

   a $8x - 6 = 3x + 9$ **b** $2(3x - 5) = 7x - 34$ **c** $\frac{x}{7} + 3 = 15$

   d $15 - 2x = 3x + 5$ **e** $5(2x + 7) = 3(4x - 5)$ **f** $8x - 2(5 + 2x) = 7x - 16$

3 If $y = 8$ what is the value of x in each of these equations?

   a $2y + 3x = 22$ **b** $4x = 3y - 8$ **c** $2(3x - y) = 5x + 7$

   d $3y - 2x = 5x - 4y$ **e** $7x + 2 = 3y - 4x$ **f** $3x - 8 = 5y$

4 Nina said, 'Multiplying my number by 4 then subtracting 5
gives the same answer as multiplying by 2 and adding 9'.

   a Call the number n and form an equation.   **b** Work out the value of Nina's number.

5 Find the value of the angle y in this triangle.

6 In these flow diagrams you get to the same finishing number
whichever route you take.

Form equations and find the starting number x.

a
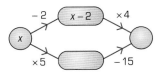

b
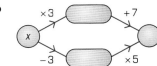

7 Work out the value of these angles.

a
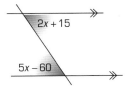

b
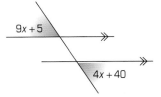

c
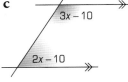

8 **a** The mean of three numbers is 18. If two of the numbers
are 10 and 20, what is the third number?

   b The mean of three expressions is $2x + 8$.
Two of the expressions are:

   What is the third expression?

This spread will show you how to:
▶▶ Use index notation.
▶▶ Use systematic trial and improvement methods.

KEYWORDS
Negative   Square root
Brackets   Solution
Positive

You can multiply positive and negative numbers in four ways:

▶ $4 \times 4 = 16$
▶ $4 \times {}^-4 = {}^-16$ There are only two different
▶ ${}^-4 \times {}^-4 = 16$ solutions: 16 and ${}^-16$.
▶ ${}^-4 \times 4 = {}^-16$

Remember:
▶ negative × negative = positive
▶ negative × positive = negative

The solution is positive when the numbers are exactly the same:

$4 \times 4 = (4)^2 = 16$ and ${}^-4 \times {}^-4 = ({}^-4)^2 = 16$

This means there are two square roots of a number:
one positive and one negative.

You use brackets to square a negative:
▶ $({}^-4)^2$ means ${}^-4 \times {}^-4 = 16$
▶ ${}^-4^2$ means ${}^-(4 \times 4) = {}^-16$

$\sqrt{16} = 4$ and $\sqrt{16} = {}^-4$

You can write $\sqrt{16} = \pm 4$

▶ For any square root there are two answers, one positive and
one negative: $\sqrt{81} = +9$ or ${}^-9 = \pm 9$

You can use square roots to solve equations involving squares.

example

Liam says that if you add 5 to the number he is thinking of,
and then square it, the answer is 36.
What two numbers could he be thinking of?
....

Let the number be n
Add 5 $n + 5$
Then square $(n + 5)^2$
The answer is 36. $(n + 5)^2 = 36$

You must use a bracket:
▶ $(n + 5)^2 = (n + 5)(n + 5)$
▶ $n + 5^2 = n + 25$

To solve the equation you undo the operations:

Square root: $n + 5 = \sqrt{36}$

There are two possible answers:

$n + 5 = 6$ or $n + 5 = {}^-6$
Subtract 5 $n = 1$ or $n = {}^-11$

The two possible numbers are 1 and ${}^-11$.

Check: $1 + 5 = 6$ $6^2 = 36$
              ${}^-11 + 5 = {}^-6$ $({}^-6)^2 = 36$

Exercise A3.3

1 Calculate these:

a 7^2 b $(^-5)^2$ c $(^-8)^2$ d 20^2

e $(^-1)^2$ f $(^-12)^2$

2 Calculate these, giving two solutions each time:

a $\sqrt{25}$ b $\sqrt{81}$ c $\sqrt{121}$ d $\sqrt{225}$

e $\sqrt{900}$ f $\sqrt{1}$

3 For each of these questions, consider only the positive solution.
Copy and complete this table.

Number	Value is between	The nearest whole number	Estimate to 1 dp	Actual answer on a calculator
$\sqrt{60}$	7 and 8	8	7.7	7.7459666
$\sqrt{20}$				
$\sqrt{39}$				
$\sqrt{48}$				
$\sqrt{7}$				
$\sqrt{56}$				
$\sqrt{84}$				
$\sqrt{110}$				

4 Work out the two possible values for x in each of these questions.

a $x^2 = 36$ b $x^2 = 100$ c $x^2 = 64$ d $x^2 + 1 = 17$

e $x^2 + 7 = 32$ f $x^2 - 10 = 26$ g $2x^2 = 50$ h $x^2 - 6 = 75$

i $x^2 + 3.5 = 52.5$

5 Work out the two possible values for x in each of these questions.

a $(x + 1)^2 = 16$ b $(x - 1)^2 = 100$ c $(x + 3)^2 = 36$ d $(x - 5)^2 = 49$

e $(x + 2)^2 = 64$ f $(x - 7)^2 = 4$

6 Solve these equations:

a $\frac{y^2}{3} = 3$ b $\frac{y^2}{6} = 6$ c $\frac{4}{y^2} = 1$ d $3 = \frac{12}{y^2}$

e $5y^2 = 45$ f $\frac{32}{y^2} = 2$ g $(5 + y)^2 = 100$ h $\frac{9}{y+1} = y + 1$

7 The area of this square is 81 m^2.

a What is the value of x?

b Why can there only be one value for x in this case?

$(x - 7)$ m

8 The area of this triangle is 50 cm^2.

a Using the formula for area of a triangle, write out an equation.

b Solve the equation to find the value of x.

$(x + 1)$ cm

$(x + 1)$ cm

43

This spread will show you how to:

▶▶ Solve problems involving direct proportion using algebraic methods.

KEYWORDS
Direct proportion
Enlargement
Ratio
Scale factor

When a photograph is enlarged, the length and width increase in the same proportion.

10 cm
7 cm
album size

15 cm
10.5 cm
large size

20 cm
14 cm
giant size

5 cm
3.5 cm
wallet size

The ratio of width : length is always the same.
This is the scale factor.

$$\frac{\text{width}}{\text{length}} = \frac{7}{10} = \frac{10.5}{15} = \frac{3.5}{5} = \frac{14}{20} = 0.7$$

The quantities are in direct proportion.

$$\frac{\text{width}}{\text{length}} = 0.7 \quad \text{so width} = 0.7 \times \text{length}$$

▶ **When two quantities increase in direct proportion, the ratio between the quantities is the same.**

You can use algebra to solve problems involving proportion.

The graph of the length and width is a straight line:

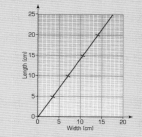

The equation of the graph is
width = 0.7 × length

example

These photographs are in direct proportion to each other.
Find the width of the larger photograph.

8 cm
14 cm

y
24.5 cm

Let the width be y.
You know that width : length is the same for each photograph.

$$\frac{y}{24.5} = \frac{8}{14}$$

$$y = \frac{8 \times 24.5}{14}$$

$$y = 8 \times 24.5 \div 14 = 14 \text{ cm}$$

Exercise A3.4

1 Find the ratio between the widths and lengths of these photographs.
Which one is not in direct proportion to the other three?

a 4 cm
10 cm

b 7 cm
17.5 cm

c 3 cm
7.5 cm

d 4 cm
1.5 cm

2 Draw a graph to confirm your answer to question 1.
Plot width on the *x*-axis and length on the *y*-axis.
Explain how the graph confirms your answer.

3 Solve these algebraic equations:
a $\frac{x}{5} = \frac{17}{10}$ **b** $\frac{y}{9} = \frac{8}{3}$ **c** $\frac{z}{3} = \frac{15}{6}$ **d** $\frac{6}{15} = \frac{a}{10}$ **e** $\frac{12}{20} = \frac{b}{15}$ **f** $\frac{1.8}{2} = \frac{c}{5}$

4 Use algebra to find the missing width or length for each of these photographs.
Each pair is in direct proportion.

a 10 cm
15 cm
x
9 cm

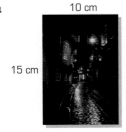

b 5 cm
y
8 cm
4.8 cm

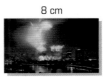

c *x*
6 cm
4 cm
2.5 cm

d 10 cm
12 cm
y
9 cm

5 Light green paint is made by mixing 5 parts of blue paint
with 12 parts of yellow paint.
Use algebra to find how many litres of:
a blue paint you would mix with 30 litres of yellow paint
b yellow paint you would mix with 12 litres of blue paint.

6 The model aircraft club make exact replicas of real aircraft.
Each dimension is in direct proportion to a real aircraft.
For the model of a Jumbo Jet the wing span is 48 cm.
The actual wing span of a Jumbo Jet is 64 m.
The actual length of the model was 56 cm.
How long is a real Jumbo Jet?

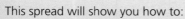
This spread will show you how to:

▶▶ Use systematic trial and improvement methods to find approximate solutions to equations.

Algebra is useful for solving problems.

When the equations are too difficult to solve you may need to estimate the answer.

> Using algebra helps you keep track of unknown values.

example

Mara thinks of a number.
She squares it.
She adds the number she first thought of.
The answer is 306.
What was her number?

....

You can use algebra to form an equation:

Let the number be	n
Square	n^2
Add n	$n^2 + n$
The answer is 306.	$n^2 + n = 306$

This is too difficult to solve so try some values.

Try $n = 10$	$10^2 = 100,$	$100 + 10 = 110$	too small
Try $n = 20$	$20^2 = 400,$	$400 + 20 = 420$	too big
Try $n = 15$	$15^2 = 225,$	$225 + 15 = 240$	too small
Try $n = 17$	$17^2 = 289,$	$289 + 17 = 306$	exactly right

She was thinking of 17.

This method is called trial and improvement.
You estimate the answer and then try to improve your estimate.
It helps to use a table for your working.

example

Use the trial and improvement method to solve the equation $2x^2 + 3x = 72.08$

Estimate	$2x^2$	$3x$	$2x^2 + 3x$	
$x = 4$	32	12	44	too small
$x = 5$	50	15	65	too small
$x = 5.5$	60.5	16.5	77	too big
$x = 5.2$	54.08	15.6	69.68	too small
$x = 5.3$	56.18	15.9	72.08	✓

So $x = 5.3$

> Using a table helps you to be systematic.

> Use a calculator to work out decimal values.

Exercise A3.5

1 If $x = 6$, work out the value of each of these expressions:

  a $x^2 - 9$ **b** $2x^2 + 7$ **c** $3x^2 + 2x$ **d** $5x^2 - 8x$

  e $10x - x^2$ **f** $2x^2 - x + 1$ **g** $25 + x - x^2$ **h** $4x^2 + 3x - 7$

2 **Best guess**

Copy and complete the table.

For each equation:

- Estimate a value for x.
- Work out the value of the expression for your value of x.
- Find the difference between your estimate and the actual value.

Work out the total difference for your four guesses.

Equation	Expression	Estimate $x = ?$	Expression value	Actual value	Difference
$3x^2 + 7 = 82$	$3x^2 + 7$			82	
$180 - 2x^2 = 52$	$180 - 2x^2$			52	
$2x^2 + 3x = 119$	$2x^2 + 3x$			119	
$3x^2 - 2x = 180$	$3x^2 - 2x$			180	
			Total:		

3 Use trial and improvement to find the exact value of x in each equation. The first one is started for you.

  a $3x^2 + 7x = 516$

Estimate	$3x^2$	$+7x$	$3x^2 + 7x$	Too small or large
$x = 10$	300	70	370	too small
$x = 15$	675	105		
$x =$				

  b $2x^2 - 5x = 777$

Estimate	$2x^2$	$-5x$	$2x^2 - 5x$	Too small or large

  c $5x^2 + 3x - 2 = 1672$

4 Use trial and improvement to find an exact value of x for these expressions:

  a $2x^2 + x = 32.68$

  b $x^2 - 3x = 41.31$

  c $3x^2 + 7x - 15 = 49.08$

  d $300 - 3x^2 = 202.53$

5 For the equation $x^2 - 6x + 8.51 = 0$, there are two positive values for x. Use trial and improvement to find both values.

This spread will show you how to:
▶▶ Use rounding to make estimates.
▶▶ Use systematic trial and improvement methods to find approximate solutions to equations.

KEYWORDS
Decimal place (dp)
Best estimate   Round
Solution   Exact

Some equations do not have exact solutions.

$x^2 = 15$ so $x = \pm\sqrt{15} = \pm 3.872983346\ldots$

You round the solution: $x = \pm 3.9$ to 1 decimal place.

You use the halfway value to decide how to round a solution.

example

$3x^2 - 4x = 17.15$
When $x = 3.1$ $3x^2 - 4x = 16.43$
When $x = 3.2$ $3x^2 - 4x = 17.92$
Which is the best estimate: 3.1 or 3.2?

Use a calculator to help you work out estimates.

⋯⋯⋯⋯⋯⋯⋯⋯⋯⋯⋯⋯⋯⋯⋯⋯⋯⋯⋯⋯⋯⋯⋯⋯⋯⋯⋯

Use the halfway value – 3.15 is halfway between 3.1 and 3.2.
When $x = 3.15$, $x^2 - 4x = 17.1675$ which is too big.
So the solution is nearer 3.1 than 3.2.
$x = 3.1$ is the best estimate.

To estimate a solution to 1 dp you work to 2 dp.

example

The area of this rectangle is 90 cm^2.
Find the value of x to 1 decimal place.

$2x$
$2x - 3$

⋯⋯⋯⋯⋯⋯⋯⋯⋯⋯⋯⋯⋯⋯⋯⋯⋯⋯⋯⋯⋯⋯⋯⋯⋯⋯⋯

Area = length × width   = $2x(2x - 3) = 90$

Using a table:

Estimate	$2x$	$2x - 3$	$2x(2x-3)$	Too big or small
$x = 5$	10	7	70	too small
$x = 6$	12	9	108	too big
$x = 5.5$	11	8	88	too small
$x = 5.6$	11.2	8.2	91.84	too big
$x = 5.55$	11.1	8.1	89.91	too small

Always check the halfway value first:
5.5 is halfway between 5 and 6
5.55 is halfway between 5.5 and 5.6

$x = 5.6$ cm (to 1 dp), as 5.55 was too small.

Exercise A3.6

1 For each of these equations:

- ▸ Find the halfway value between the two estimates and substitute it into the equation.
- ▸ Decide which is the best estimate.
  Give a reason for your answer.

  a $x^2 + 4x = 19$; $x = 2$ and $x = 3$
  b $2x^2 + 3x = 120$; $x = 6$ and $x = 7$
  c $5x^2 - 8x = 200$; $x = 7$ and $x = 8$
  d $x^2 + 3x = 90$; $x = 8.1$ and $x = 8.2$

2 Use trial and improvement to find the value of x in these equations to 1 dp.
Copy and complete the tables to help you.

a $x^2 + 5x = 114$

x	x^2	$5x$	$x^2 + 5x$	Too big/small

b $2x^2 - 7x = 13$

x^2	$2x^2$	$7x$	$2x^2 - 7x$	Too big/small

3 The area of this rectangle is 108 cm^2.
Copy and complete the trial and improvement table to find an estimate (to 1 dp) for the value of x.

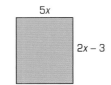

x	$5x$	$2x - 3$	$5x(2x - 3)$	Too big/small

4 The table shows values of x and y for the equation $y = x^2 + 3x - 5$.
  a Copy and complete the table:

x	$^-2$	$^-1$	0	1	2	3	4
y				$^-1$	5	13	23

The value of y is 0 for a value of x somewhere between 1 and 2.
  b Use trial and improvement to find the value of x, to 1 dp, that gives the value of y closest to 0.
  c Plot the coordinates on a graph.
  Join them with a smooth curve.
  Does the graph help you solve the question in part **b**?

5 Use trial and improvement to find the value of x, to 1 dp, that gives the value of y closest to 0 in each case [$x > 0$].
  a $y = 3x^2 - 2x - 17$
  b $y = x^2 - 7x + 3$

You should know how to ...

1 Construct and solve linear equations with integer coefficients using an appropriate method.

Check out

1 a If $y = 4$ and $z = 6$, find the value of x in each of these:

  i $5x - 3y = 38$

  ii $2y + 4x = 3z$

b Use this triangle to form an equation and find the value of y.

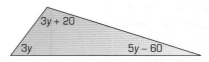

c The widths and lengths of these rectangles are in direct proportion. Find the value of x.

2 Use trial and improvement where a more efficient method is not obvious.

2 a Copy and complete the table, using trial and improvement to find the exact value of x in this equation:

$$2x^2 - 3x = 106.92$$

Estimate	$2x^2$	$3x$	$2x^2 - 3x$	Too small / too big
8				

b The area of this rectangle is 72 cm². Use trial and improvement to find the value of x to 1 dp.

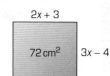

Geometrical reasoning and construction

This unit will show you how to:

▶▶ Distinguish between conventions, definitions and derived properties.

▶▶ Explain how to find, calculate and use:
  – sums of the interior and exterior angles of quadrilaterals, pentagons and hexagons
  – interior and exterior angles of polygons.

▶▶ Solve problems using properties of angles, of parallel and intersecting lines, and of triangles and other polygons.

▶▶ Know the definitions of a circle and its parts.

▶▶ Explain why inscribed regular polygons can be constructed by equal divisions of a circle.

▶▶ Use straight edge and compasses to construct a triangle, given right angle, hypotenuse and side (RHS).

▶▶ Use ICT to explore constructions of triangles and other 2-D shapes.

▶▶ Find the locus of a point that moves according to a simple rule.

▶▶ Explore connections in shape and space across a range of contexts.

▶▶ Represent problems and synthesise information in geometric form.

▶▶ Present a concise, reasoned argument.

Engineers use properties of shapes to design structures.

Before you start

You should know how to ...

1 Use angle facts:
  ▶ The angles in a triangle add up to 180°.
  ▶ The exterior angle of a triangle equals the sum of the two opposite interior angles.
  ▶ The angles on a straight line add to 180°.

2 Recognise special triangles and quadrilaterals and describe their properties.

Check in

1 Find the missing angles.

 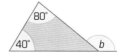

2 Name these shapes and describe their properties.

a
b

This spread will show you how to:

▶▶ Distinguish between conventions, definitions and derived properties.

KEYWORDS
Convention   Degree
Definition   Sketch
Derived property

▶ A **convention** is an agreed way of describing a situation.

This is the line AB.

This is the angle BAC, or ∠BAC.

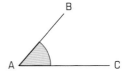

There are conventions for:
equal lengths and angles ...

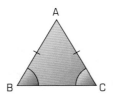

AB = AC
∠B = ∠C

... and parallel lines.

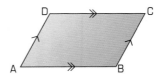

BC is parallel to AD.
AB is parallel to DC.

▶ A **definition** is a set of conditions needed to specify a particular object.

A **polygon** is a closed shape with straight sides.

An **equilateral triangle** is a triangle with three equal sides.

A **degree** is a unit for measuring angles. One complete rotation is divided into 360 degrees.

▶ A **derived property** follows on from a definition.

The opposite sides of a parallelogram have equal lengths.

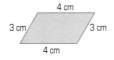

4 cm
3 cm   3 cm
4 cm

Each angle in a square measures 90°.

Half a complete rotation is 180°.

Exercise S1.1

1 Sketch a diagram to show each of these labelling conventions:
   a the line CD
   b the angle CAB
   c the triangle ABC
   d the parallelogram ABCD
   e the isosceles triangle XYZ where XY = YZ
   f the parallelogram EFGH where EF = HG
   g the trapezium ABCD where AB is parallel to DC
   h the equilateral triangle ABC
   i the trapezium ABCD where ∠A = ∠B = 90° and BC is parallel to AD
   j the triangle ABC with AC extended to D.

2 Copy and complete these definitions:
   a An isosceles triangle has two pairs of equal ___ and ___.
   b A ___ is a quadrilateral with one pair of parallel sides.
   c A ___ shape has equal angles and equal sides.
   d The vertex of a shape is ___.
   e A quadrilateral has ___ sides and ___.
   f A ___ has six sides.

3 Explain why these definitions are incomplete.
   Rewrite the sentences to make them into definitions.
   a A square has four sides of equal length.
   b A parallelogram has opposite sides of equal length.
   c A trapezium has two parallel sides.
   d A regular hexagon has six sides.
   e An equilateral triangle has three sides.

4 On a copy of this diagram, mark a pair of corresponding angles and a pair of alternate angles.

5 Copy and complete these sentences that describe some derived properties.
   a The angles of a triangle add up to ___.
   b Opposite angles of a parallelogram are ___.
   c A square has diagonals that are ___ in length.
   d Each interior angle in an equilateral triangle is ___.
   e Each interior angle in a square is ___.
   f Opposite sides of a parallelogram are ___ and ___.
   g The sides of a rhombus are ___.

6 Use a triangular piece of paper.

   Fold the corners of your triangular piece of paper to a common point on the base. What does this show?

   Does the result always work? Why? Explain and justify your reasoning.

This spread will show you how to:
▶▶ Explain how to find, calculate and use the sums of the interior and exterior angles of quadrilaterals, pentagons and hexagons.

KEYWORDS
Polygon   Vertex
Interior angle   Vertices
Exterior angle   Quadrilateral
Pentagon   Hexagon

Here is a polygon.

The blue angles are **interior** angles.
The red angles are **exterior** angles.

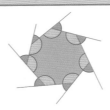

Interior angles

You can split a quadrilateral into two triangles:

The total angle sum is $2 \times 180° = 360°$

The angles of a triangle add up to 180°.

You can split ...

... a pentagon into three triangles:

... a hexagon into four triangles:

... an n-sided polygon into $(n-2)$ triangles:

The angle sum is $3 \times 180° = 540°$

The angle sum is $4 \times 180° = 720°$

The angle sum is $(n-2) \times 180°$

▶ The interior angle sum of a polygon is $(n-2) \times 180°$.

Exterior angles

At each vertex of a polygon:
interior angle + exterior angle = 180°

For a polygon with n vertices, the sum of the interior and exterior angles is $n \times 180°$.

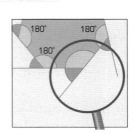

The angles on a straight line add up to 180°.

A vertex is a corner of a shape.

The interior angles add up to $(n-2) \times 180°$.
So the exterior angles add up to:
$$n \times 180° - (n-2) \times 180° = (n \times 180°) - (n \times 180°) + (2 \times 180°)$$
$$= 2 \times 180°$$
$$= 360°$$

This is a **derived property** as it follows from known facts.

▶ The exterior angle sum of a polygon is always 360°.

Exercise S1.2

1 The diagram shows a quadrilateral.

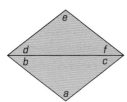

Copy and complete:

$a + b + c =$ ____

$d + e + f =$ ____

The angle sum of a quadrilateral = ____

2 Use this diagram to prove that the sum of the three angles of a triangle is 180°.

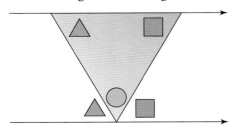

Hint: First look at the interior angles, then the angles on the straight line.

3 For this pentagon:

Copy and complete:

a You can split the pentagon into ____ triangles from one vertex.

b Each triangle has an angle sum of ____°.

c The angle sum of the pentagon is ____ × ____°.

4 Find the sum of the interior angles of each of these polygons.

a

b

c

5 Find the sum of the interior angles of:
a an octagon
b a decagon
c a 30-sided polygon
d a 100-sided polygon.

6 Find the sum of the exterior angles of each polygon in question 5.

7 A regular polygon has equal sides and equal angles. This tessellation is made from regular octagons and regular quadrilaterals.

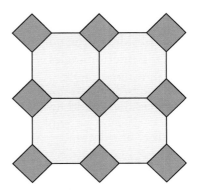

Draw a tessellation that uses a regular polygon and one other shape.

8 Use your knowledge of angles to explain why the tessellation in question 7 works.

Angles in regular polygons

This spread will show you how to:

▶▶ Explain how to find, calculate and use the interior and exterior angles of regular polygons.

KEYWORDS

Regular polygon   Sum

Interior angle

Exterior angle

In a **regular polygon** all the interior angles are equal.

The sum of interior angles, $S = (n - 2) \times 180°$

A regular hexagon has six sides, so $n = 6$.
$$S = (n - 2) \times 180°$$
$$= (6 - 2) \times 180°$$
$$= 720°$$

n is the number of vertices.

Each interior angle $= \dfrac{720°}{6} = 120°$

You can work out the exterior angle by using either of two methods:

At any vertex of a polygon,
interior angle + exterior angle = 180°

So exterior angle of a regular
hexagon = 180° − 120° = 60°.

The exterior angle sum of a polygon is 360°.

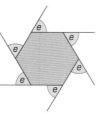

So exterior angle of a regular hexagon $= \dfrac{360°}{6} = 60°$.

You can use these techniques to solve problems involving regular polygons.

example

The exterior angle of a regular polygon is 15°. Find:

a the number of sides of the polygon
b the sum of the interior angles
c the size of an interior angle.

....

a Exterior angles always add up to 360°.
So the number of sides $= \dfrac{360°}{15°} = 24$

b The sum of the interior angles is $(n - 2) \times 180°$.
$n = 24$, so sum of interior angles $= 22 \times 180°$
$$= 3960°$$

c Exterior angle = 15°
Interior angle = 180° − 15°
$$= 165°$$

You should learn these
interior angles:

Shape	Interior angle
Equilateral triangle	60°
Square	90°
Regular hexagon	120°

Exercise S1.3

1 The diagram shows a pentagon.

Copy and complete:

> You can split a regular pentagon into 3 triangles.
> The sum of the interior angles is
> $3 \times 180° = $ ___.
> Each interior angle is ___ ÷ 5 = ___.
> Each exterior angle is $180° - $ ___ = ___
> or $\dfrac{360°}{5} = $ ___.

2 The diagram shows an octagon.

Copy and complete:

> You can split a regular octagon into ___ triangles.
> The sum of the interior angles is
> ___ $\times 180° = $ ___.
> Each interior angle is ___ ÷ 8 = ___.
>
> Each exterior angle is $180° - $ ___ = ___
> or $\dfrac{360°}{8} = $ ___.

3 Find the sum of the interior angles of each of these polygons.

a   **b**   **c**

 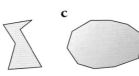

4 Use your answers to question 2 to work out the interior angle of each of these regular polygons:
  a regular hexagon
  b regular nonagon
  c regular decagon.

5 Find the exterior angle of a:
  a regular octagon
  b regular hexagon
  c regular decagon
  d regular nonagon.

6 Use your answers to question 5 to find the interior angle of a:
  a regular octagon
  b regular hexagon
  c regular decagon
  d regular nonagon.

7 In questions 4 and 6 you have used two methods to find the interior angle of a regular polygon. Which do you prefer? Give a reason for your choice.

8 **Investigation**
What is the maximum number of right angles you can have in an octagon?

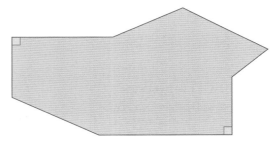

What about other polygons?

S1.4 Lines and angles

This spread will show you how to:

▶▶ Solve problems using properties of angles, of parallel and intersecting lines, and of triangles and other polygons.

KEYWORDS
Vertically opposite
Alternate   Corresponding
Parallel

You should remember these types of angles:

a and *b* are **vertically opposite** angles.

c and *d* are **corresponding** angles.

e and *f* are **alternate** angles.

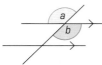

Vertically opposite angles are equal.

Corresponding angles are equal.

Alternate angles are equal.

You can use the properties of intersecting parallel lines to solve problems.

example

Find the value of each of the missing angles.

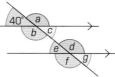

$a = 180° - 40° = 140°$ (angles on a straight line)
$b = 140°$ (vertically opposite a)
$c = 40°$ (vertically opposite $40°$)
$d = 40°$ (corresponding to a)
$e = 40°$, $f = 140°$, $g = 40°$

You can also use the properties of shapes to solve problems.
You should explain your working, using diagrams where possible.

Diagrams help you follow your working.

example

The diagram shows a quadrilateral ABCD.
Find:

a ∠BDC   **b** ∠BAD   **c** ∠ADC
d What is the name of the shape ABCD?

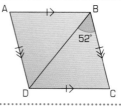

a △BCD is isosceles
   ⇒ ∠BDC = ∠DBC
         = 52°

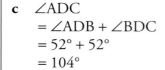

c ∠ADC
   = ∠ADB + ∠BDC
   = 52° + 52°
   = 104°

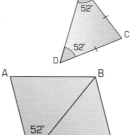

b ∠ADB = ∠DBC
         = 52°   (alternate angles)
   △BAD is isosceles
   ⇒ ∠ABD = ∠ADB
         = 52°
   ∠BAD = 180° − (52° + 52°)
         = 76°   (angles in a triangle)

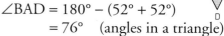

d The sides are all equal and opposite angles are equal.
   So the shape is a rhombus.

Exercise S1.4

1 Sketch these diagrams on squared paper and name them.
  a A shape with four straight sides, four right angles and two pairs of parallel sides.
  b A shape with four straight sides, no right angles and two pairs of parallel sides.
  c A shape with four straight sides, four right angles and four sides of equal length.

2 **a** Sketch an isosceles triangle PQR.
  b Mark the equal angles and sides and mark the angle PQR.
  c Extend the side PQ to S and mark the exterior angle of the triangle.

3 Find the angles marked with letters in these diagrams.

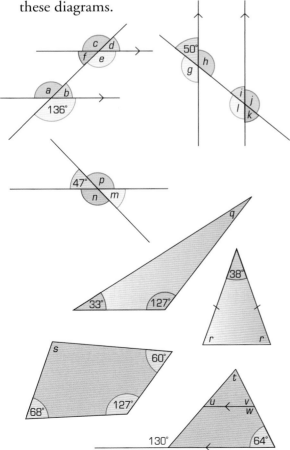

4 **a** Draw a pair of parallel lines and cross them with an intersecting line.
  b Label the parallel lines with arrowheads.
  c Label any pair of alternate angles with the letter *a*.
  d Label any pair of corresponding angles with the letter *c*.
  e Label any pair of vertically opposite angles with the letter *v*.

5 The diagram shows a rectangle ABEF resting on an equilateral triangle BCD. DE is a straight line passing through B. Show that triangle ABC is isosceles.

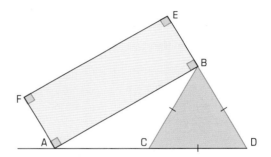

6 **Investigation**
Change the equilateral triangle in question 5 to an isosceles triangle. Choose various values for the size of ∠CBD. Investigate the change to triangle ABC.

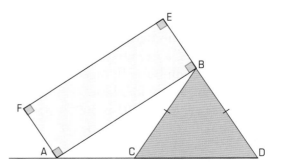

This spread will show you how to:

▶▶ Solve problems using properties of angles, of parallel and intersecting lines, and of triangles and other polygons.

KEYWORDS

Triangle	Quadrilateral
Rhombus	Kite

You should already know that:

▶ The exterior angle of a triangle equals the sum of the two opposite interior angles.

In the diagram, $a + b = d$.

example

Find the angle marked x.

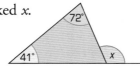

$x = 41° + 72° = 113°$
(exterior angle equals sum of opposite interior angles.)

You need to know the angle properties of special quadrilaterals.

In a **parallelogram** the opposite angles are equal.

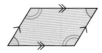

In a **rhombus** the diagonals bisect each other at right angles.

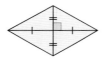

A **kite** has a single line of symmetry. The other diagonal splits the kite into two isosceles triangles.

You can use these properties to solve problems.

example

The diagram shows an arrangement of three identical tiles.
a What shape are the tiles?
b Find the interior angles of the tiles.

a The shape has equal-length sides as they all fit together exactly.
Opposite sides are parallel.
There are no right angles.
It must be a rhombus.
b These rhombuses fit together so the angles must add to 360°.
The angles are identical.
They are each 120° as 360° ÷ 3 = 120.
The other angles must be 60°.

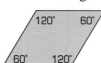

$360 - (2 \times 120) = 120$
$120 \div 2 = 60$

Exercise S1.5

1 Find the angles marked with letters in these diagrams.

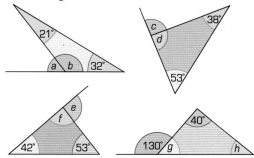

2 Sketch each of these shapes, marking any equal angles, equal sides and lines of symmetry:
  a rhombus
  b regular octagon
  c isosceles triangle
  d equilateral triangle
  e kite
  f trapezium
  g scalene acute-angled triangle
  h regular pentagon
  i irregular quadrilateral
  j isosceles trapezium.

3 Find the angles marked with letters in these quadrilaterals.

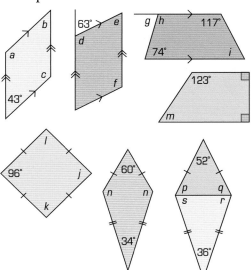

4 Name each of the quadrilaterals in question 3.

5 Write a LOGO program to draw each of these polygons with a side of 5 cm. The first one is started for you.
  a Regular pentagon
    FORWARD 50
    TURN RIGHT 72
    FORWARD 50
    TURN RIGHT 72
    FORWARD__
  b Equilateral triangle
  c Regular hexagon
  d Regular decagon
  e Square

6 The diagram shows a rhombus.

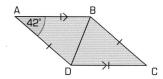

Find these angles:
  a ∠ABD
  b ∠ADB
  c ∠BDC
  d ∠BCD
  e ∠DBC

7 PQRS is an isosceles trapezium.

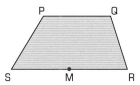

  a M is the midpoint of RS.
    If ∠QRS is 55°, calculate the size of the other angles.
  b Repeat with different values for ∠QRS.

S1.6 Circle properties

This spread will show you how to:
- ▶▶ Know the definition of a circle and the names of its parts.
- ▶▶ Explain why inscribed regular polygons can be constructed using equal divisions of a circle.

KEYWORDS

Chord	Semicircle
Centre	Equidistant
Circumference	Tangent
Radius	Segment
Diameter	Arc
Sector	

A circle is a set of points with an interesting property ...

A and B are exactly the same distance from O.

The points A to F are the same distance from O.

All of these points are the same distance from O.

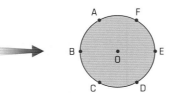

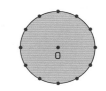

▶ A circle is a set of points **equidistant** from its centre.

Equidistant means 'the same distance from a fixed point'.

▶ The **circumference** (C) is the distance around the circle.
▶ The **radius** (r) is the distance from the centre to the circumference.
▶ The **diameter** (d) is the distance across the circle.
▶ The diameter divides the circle into two **semicircles**.

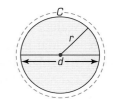

You also need to know these parts of a circle.

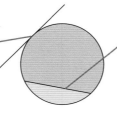

A **tangent** is a line that touches the circle at a single point.

A **chord** is a line joining two points on the circumference. It divides the circle into two **segments**.

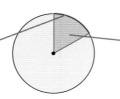

An **arc** is a part of the circumference.

A **sector** is a region bounded by an arc and two radii.

Radii is the plural form of radius.

Exercise S1.6

1 Draw a circle with radius 4 cm. Label the centre, the radius and the circumference.

2 **a** Draw an arc with radius 4 cm. On this arc, draw a sector with angle 60°.
   b Draw an arc with radius 5 cm. On this arc draw a sector with angle 120°.
   c Draw a circle with radius 5 cm. On this circle draw and label:
   - **i**   an arc
   - **ii** a sector
   - **iii** a radius
   - **iv** a chord
   - **v**   a segment.

3 You can use a circle to construct a regular hexagon.
Follow these steps:
   a Using compasses, draw a circle of radius 5 cm.

   b Without changing your compass setting, put your compass point on the circle and mark off points. Put your compass point on each mark you have just made to make the next mark.

   c When you have got back to the start, join the points up with chords. This will give you a regular hexagon.

4 Add these lines to the hexagon you drew in question 3.

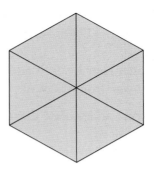

Explain how you know that a regular hexagon is made up of six equilateral triangles.

5 Construct a regular pentagon by following these steps:
   a Work out this angle first.

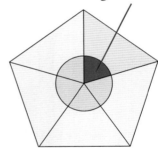

   b Draw a circle of radius 6 cm. Draw in the radius.
   c Measure the angle you found in part **a** from the radius.
   Mark the point on the circumference. Join the centre to this point.
   d Repeat part **c** from your new radius.
   e Repeat again until you return to the start.
   You should now be able to produce the pentagon.

6 Which other regular polygons could you construct by equal divisions of a circle?

Constructing right-angled triangles

This spread will show you how to:
▶▶ Use a straight edge and compasses to construct a triangle, given right angle, hypotenuse and side.

KEYWORDS
Construction lines
Hypotenuse   Arc
Scale drawing Perpendicular
Compasses

A window cleaner rests his 9 ft ladder against a wall.
Its base is 4 ft away from the wall.

You can construct an accurate diagram using a **scale drawing**.

1 First, draw a rough sketch.

The ladder is the **hypotenuse** in a right-angled triangle.

2 Decide on the **scale**.
1 cm to 1 ft will give sensible lengths.
Scale:
1 cm represents 1 ft.

3 Draw a line 4 cm long to represent the **base**.

A   4 cm   B

4 To draw a **perpendicular** at B, extend your compasses to about 2 cm. With the centre at B, make arcs as shown.

A   X   B   Y

7 Join the points A, B and C to complete the right-angled triangle.

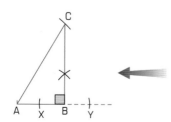
A   X   B   Y

Do not rub out your construction lines.

6 For the ladder, set your compasses to 9 cm. With centre A, make a mark on the line representing the wall.
Label it C.

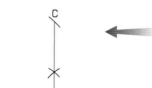

5 Keeping your compasses at the same setting, make arcs above and below B with the centre at X. Repeat with Y as shown.
Join the points where the arcs cross to make the wall.

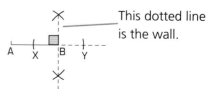
This dotted line is the wall.

Exercise S1.7

1 Construct the scale drawing described on page 64.
Use your diagram to measure:
a the angle between the ladder and the ground
b how far up the wall the ladder reaches.

2 Construct these triangles.

a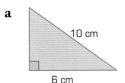

b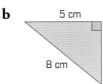

c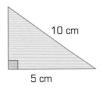

3 Construct these triangles.

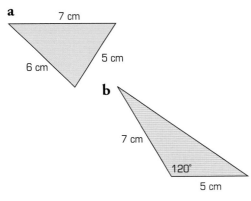

d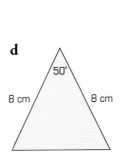

4 A window cleaner has a 12 metre ladder. Kings Castle is surrounded by a 4 metre moat. Use a scale drawing to investigate the maximum height of windows that can be cleaned.

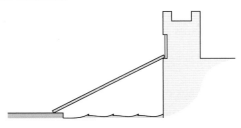

5 Construct these triangles.

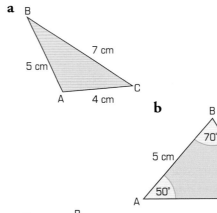

c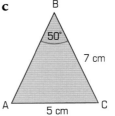

6 Can you construct triangles with these measurements?
Construct each triangle if possible. If not possible, give a reason.
a $\angle BAC = 50°$, $\angle ABC = 60°$, $\angle ACB = 70°$
b $BC = 6$ cm, $AC = 4$ cm, $AB = 2$ cm
c $AC = 8$ cm, $BC = 6$ cm, $\angle ACB = 40°$
d $BC = 6$ cm, $AC = 10$ cm, $\angle ABC = 80°$

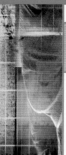

S1.8 Drawing loci

This spread will show you how to:
▶▶ Find the locus of a point that moves according to a simple rule.

KEYWORDS
Locus   Circle
Perpendicular bisector

Pizza Rapido will deliver within a 6 km radius of the shop.

You can show the boundary of the delivery area as a locus.

The locus is a circle.

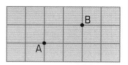

Scale:
1 cm represents 1 km.

▶ The locus of a set of points that are a given distance from a fixed point is a **circle**.

If the conditions specify **two** fixed points, the locus is different.

example

A and B are two fixed points on a grid.
Mark on the grid the locus of all points that are equidistant from A and B.

All of these points are equidistant from A and B.

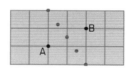

If you join them up you get the locus.

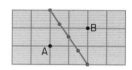

▶ The locus of a set of points that are a given distance from two fixed points is the **perpendicular bisector** of the line joining the two points.

You may have to construct a locus accurately.

example

Beaconford is 5 km east of Allersham.
A new road is to be built so that it is equidistant from Allersham and Beaconford.
Accurately construct a diagram to show the new road using a scale of 2 cm to 1 km.

Allersham   Beaconford

1 Draw a horizontal straight line AB, 10 cm long.

2 Draw arcs from A.

3 Repeat with the point at B, and join the arcs.

A •————————• B

A •————————• B

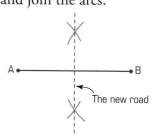

A •————————• B

The new road

Exercise S1.8

1 Copy the diagrams and draw the locus of all points which are the same distance from the marked points.

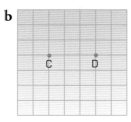

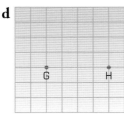

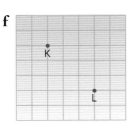

g

M

N

2 Pizza Rapido delivers within a 6 km radius. Pizza Quik delivers within a 4 km radius. The shops are 5 km apart. Draw a scale diagram to show the region which can receive pizzas from both outlets.

Start by drawing R and Q 5 km apart. A scale of 1 cm to 1 km is suitable.

3 Use a straight edge and compasses to draw the locus of all the points that are the same distance from these pairs of points.

a

A • 8 cm • B

b C • 10 cm • D

c E • 12 cm • F

4 A spider is motionless.
A fly is buzzing so that it is always 8 cm from the spider.

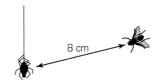

8 cm

Describe the locus of the fly.

Hint:
The fly moves in 3-D space!

5 Describe two other situations which would give a locus of the same shape as the one in question 4.

This spread will show you how to:

▶▶ Find the locus of a point that moves according to a simple rule.

KEYWORDS
Locus   Bisector

Isobel is making loci using counters on a table.

She places counters that are equidistant from the opposite edges AB and DC.

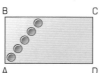

The locus is a line parallel to AB and DC, and exactly halfway between them.

Isobel then places counters that are equidistant from the adjacent edges AB and AD.

The locus is the **angle bisector** of AB and AD.

An angle bisector cuts the angle in half.

▶ The locus of a set of points that is at a given distance from two fixed lines is the **angle bisector** of the lines.

example

Construct the locus of points that are equidistant from the lines AX and AY.

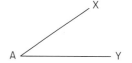

1 Using compasses, draw an arc from A that cuts both lines.

2 Draw arcs from P and Q as shown, using the same compass setting.

3 Join AB. This line bisects the angle XAY.

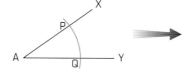 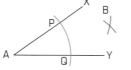 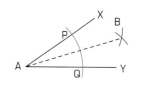

The locus of a moving object is harder to describe.
It helps to visualise the path that is traced out.

example

A square is rolled along a straight line. Sketch the locus of the vertex A.

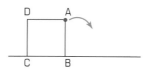

The path of the vertex A would be a curve as shown.

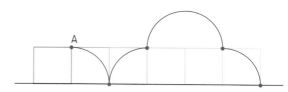

Exercise S1.9

1 Copy the diagrams and draw the locus of all points which are the same distance from both lines.

a

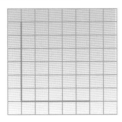

b

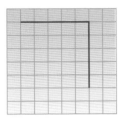

c

d

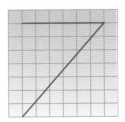

2 On plain paper, draw three diagrams with two straight lines similar to those in question 1.
Construct the locus of all points which are the same distance from both lines for each diagram.

3 Using only compasses, a ruler and a pencil:
  a Construct an angle of 90°.
  b Construct an angle of 45°.

4 Using only compasses, a ruler and a pencil:
  a Construct an equilateral triangle.
  b Bisect each angle.
  c Describe the different shapes made inside the triangle by the angle bisector.

5 Look at the last example on page 68. Copy and continue the locus of the point A as the square is rolled along a straight line.

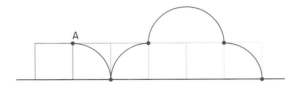

6 An isosceles right-angled triangle is rolled along a straight line:

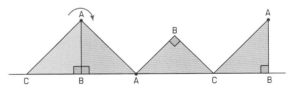

Copy and continue the locus of the vertex A.

7 Investigate the locus of a fixed point A for other regular polygons rolled along a straight line.

Summary

You should know how to ...

1 Solve problems using properties of angles, of parallel and intersecting lines, and of triangles and other polygons.

Check out

1 Here is a triangle ABC.

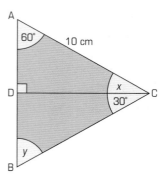

Copy and complete these statements about the triangle:

a The angle x is ____ because ____.

b The angle y is ____ because ____.

c The length AB is ____ because ____.

d The length BC is ____ because ____.

e Paul says that the area of triangle ABC is 100 cm². Gill says it must be less than 100 cm².

Who is correct? Explain your answer.

2 Present a concise, reasoned argument using diagrams and related text.

2 Is it possible to construct a triangle with these measurements:

▶ AB = 3 cm

▶ BC = 3 cm

▶ AC = 8 cm?

Explain and justify your answer.
Include at least one diagram.

Handling data

This unit will show you how to:

▶▶ Suggest a problem to explore using statistical methods, frame questions and raise conjectures.

▶▶ Discuss how data relate to a problem and identity possible sources.

▶▶ Design a survey to capture the necessary data.

▶▶ Determine the sample size and degree of accuracy needed.

▶▶ Design, trial, and if necessary refine data collection sheets.

▶▶ Select suitable graphical representation to progress an enquiry.

▶▶ Find summary values that represent the raw data, and select the statistics most appropriate to the problem.

▶▶ Interpret graphs and diagrams and draw inferences.

▶▶ Compare two or more distributions and make inferences.

▶▶ Communicate interpretations and results of a statistical enquiry.

▶▶ Represent problems and synthesise information in graphical form.

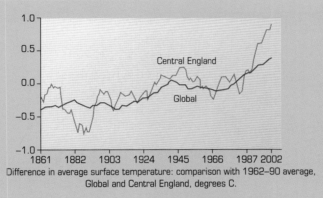

Difference in average surface temperature: comparison with 1962–90 average,
Global and Central England, degrees C.

You can see the relationship between sets of data on a graph.

Before you start

You should know how to ...

1 Calculate the mean, median, mode and range of a set of raw data.

2 Draw graphs and charts to represent data.

Check in

1 Write these numbers in size order, starting with the smallest:

a 1, 7, 8, 6, 2, 3, 7

b 1.2, 1.4, 1.3, 1.9, 0.6, 1.5, 0.8, 1.2

2 The table shows the favourite colours of 8C.

Colour	Red	Blue	Black	Purple
Number	6	9	5	10

Draw a suitable chart to represent this data.
Give a reason for your choice of diagram.

This spread will show you how to:
▶▶ Suggest a problem to explore using statistical methods.
▶▶ Discuss how data relate to a problem and identify possible sources.
▶▶ Determine sample size and degree of accuracy needed.

KEYWORDS
Specify   Interpret
Represent   Primary data
Sample

There are four main stages in a statistical enquiry.

1 Decide the problem you want to investigate then plan it carefully.

Wayne is doing a project on reaction times.

> My main hypothesis is that boys have faster reaction times than girls.

2 Choose the most appropriate data.

Liska is using both primary and secondary data for her project on mobile phones.

> I looked up national data on the internet, and also asked 40 students in my school. I wanted it to be accurate, so I chose a large sample size.

4 Interpret the results of your survey.

Kelly is interpreting her data on distance travelled to school. The mean distance is 0.75 km.

> This suggests that my hypothesis that most students live more than 1 km from school may be untrue.

1
Specify the problem and plan

2
Collect data from a variety of sources

3
Process and represent data

4
Interpret and discuss data

3 Analyse your data and represent it effectively.

David is choosing an appropriate graph for his project on height.

> I'm drawing a scatter graph because I've got two variables, height and weight.

Your interpretation may lead to new questions. You start the handling data cycle again.

Exercise D1.1

1 Penny decides to find out how accurately people can throw. She tests people's accuracy in throwing a beanbag.

   a Make a list of the factors that influence people's accuracy in throwing a beanbag, and write a hypothesis that Penny could test with an experiment.

   b Design an experiment and data collection sheet to test this hypothesis. Explain what data to collect, describe the accuracy needed in recording the data, and explain a good sample size to use.

> ▸ Too small a sample may give unrepresentative results.
> ▸ A large sample may be expensive and take a long time.

2 Jenny investigates people's attitudes to organic fruit and vegetables. She decides to carry out a survey.

   a Suggest some hypotheses that Jenny could test in her survey.

   b Design a questionnaire for a survey, based on these hypotheses.

   c Explain how Jenny could carry out a pilot survey.

   d Explain how Jenny should select the people to take part in her survey.

3 Jodie was asked to write a project with the title 'Teenagers and mobile phones'.

   a Think of three different questions that she could investigate for the project, and write a hypothesis for each one.

   b Explain how she could collect the information needed to test each of the hypotheses. Remember to include:
   ▸ Whether she would use primary or secondary data.
   ▸ How and where to collect the data.
   ▸ How big a sample size she would need.

4 A group of students was doing a project to find out whether people in the UK had healthy lifestyles.

   a One question was:

   What is your age (in years)?
   ☐ 0–15   ☐ 15–30   ☐ 30–45   ☐ 45+

   Explain why this question needs to be changed.

   b Another question was:

   Do you eat lots of vegetables?
   ☐ None   ☐ A few   ☐ Lots

   Explain why this question needs to be changed, and write a better version.

   c The students discuss who they will give their questionnaire to.
   One student said:

   > 'Let's ask everybody in our class to fill in a copy.'

   Give one advantage and one disadvantage of this suggestion.

5 Gill and Paul see a magazine article that says: '60% of teenagers now own a mobile phone'. They decide to test this claim.

   a Gill says: 'We could ask 10 people in Year 9'.
   Give two reasons why Gill's method might not give reliable results.

   b Paul says: 'We could go into town, and watch people walking past the shops. We could make a note of how many of the teenagers are carrying mobile phones'.
   Give two reasons why Paul's method might not give reliable results.

This spread will show you how to:
▶▶ Find summary values that represent the raw data, and select statistics most appropriate to the problem.

▶ An average is a single value that summarises a set of data. The **mean, median** and **mode** are three different averages.

▶ You can use the **range** to show the spread of the data.

example

The numbers of tracks on ten different CDs were:
12, 16, 12, 10, 24, 13, 11, 8, 12 and 10. Find:

a the mean   **b** the median   **c** the mode   **d** the range.

a Find the total number of tracks:
   Divide by the number of CDs: $128 \div 10 = 12.8$
   The mean number of tracks is 12.8.
b Put the numbers in order: 8, 10, 10, 11, 12, 12, 12, 13, 16, 24
   Find the middle value: ↑
   The median is 12.
c Find the most common number of tracks: 12 occurs most often
   The mode is 12.
d The range = $24 - 8 = 16$

You should decide which average is the best to use for your data.

▶ The **mode** is easy to work out, but does not always give a representative or unique value.
▶ The **mean** uses all of the data, but can be distorted by a single **extreme** value (a value that is either much larger or much smaller than the rest of the data).
▶ The **median** is unaffected by extreme values, but does not use all the data.

With continuous data, or grouped discrete data, you can work out the **modal class** instead.

A data set containing extreme values is said to be **skewed**.

example

Find the mean, median and mode of these sets of numbers. Comment on your answers.

a   1, 1, 3, 5, 10   **b** 1, 1, 2, 3, 3, 4, 35

a   Mean = $20 \div 5 = 4$, median = 3, mode = 1
   The mode is not representative of the data.
   Either the mean or the median could be used to represent the data.
b   Mean = $49 \div 7 = 7$, median = 3, mode = 1 and 3
   There are two modes, and the mean is not representative of the data.
   The median should be used here.

Exercise D1.2

1 The table shows the amount of money Joe paid for his lunch each day one week.

Monday	Tuesday	Wednesday	Thursday	Friday
£3.45	£4.50	£3.75	£4.25	£3.90

Joe says: 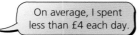 On average, I spent less than £4 each day.

Is Joe correct? Explain your answer.

2 To qualify for the final of a gymnastics competition, Polly needs a mean score of at least 9.0 points over four rounds. Her points for the first three rounds are shown in the table.

	Round 1	Round 2	Round 3	Round 4
Points	9.2	8.6	8.8	

How many points does Polly need to score in Round 4?

3 **a** Jodie played four holes of Crazy Golf. She scored 6 on each of the first three holes, and 2 on the last one. What was Jodie's mean score over four holes?
b Katryn only played two holes. Her mean score was 7, and the range of her scores was 4. What were Katryn's scores on each hole?
c Laura played three holes of Crazy Golf. She also had a mean score of 7, with a range of 4. What could Laura's scores have been on each hole? Explain your reasoning.

4 Sammy has three number cards. You cannot see the numbers, but you can work them out from the clue given.

Clue
The mean is 7
The mode is 8

What are the numbers on Sammy's cards?

5 Find sets of whole numbers that meet these conditions:
a three numbers with a range of 6, a median of 8 and a mean of 8
b three numbers with a mode of 4, a median of 4 and a mean of 5
c four numbers with a mode of 4, a median of 4, a mean of 4 and a range of 6.

6
> You can use an assumed mean to make calculations easier:
> ▸ Subtract the assumed mean from each value.
> ▸ Find the mean of the new values.
> ▸ Add on the assumed mean to give the actual mean.

Calculate the mean of each of these sets of numbers, using an appropriate **assumed mean**. Explain your working carefully.
a 32, 36, 38, 31, 30, 31, 32, 35, 34, 38
b 102, 98, 97, 96, 105, 101, 104, 99, 94, 94
c 5.26, 5.39, 5.01, 5.47, 5.66

7 The table shows the scores achieved by 20 students who played a game.

Score	0	1	2	3	4	5
Frequency	4	9	4	2	0	1

Calculate the mean score for the 20 students.

8 For each data set, state which measure of average you would use. Give a reason for your choice.
a 7, 9, 14, 26, 29, 33, 37
b 1, 1, 3, 3, 4, 24
c 1, 31, 33, 35

This spread will show you how to:
- Select, construct and modify, on paper and using ICT, suitable graphical representation to progress an enquiry.
- Identify key features present in the data.

KEYWORDS:
Graph
Scatter graph
Line graph
Frequency diagram
Stem-and-leaf diagram
Bar chart
Pie chart

When you have collected some data, you should find the best way to display it.

If your data consists of single values, then you could choose one of these charts.

A **pie chart** compares the size of each category with the total.

Favourite foods of class 9F

crisps

chocolate

A **bar chart** compares categories with each other.

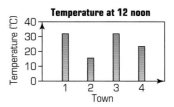
Temperature at 12 noon

A **frequency diagram** is useful for continuous data.

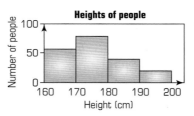
Heights of people

A **stem-and-leaf diagram** shows the shape of a distribution.

Lengths of fish in cm						
30	3	6				
20	0	1	3	5	7	9
10	2	4	5			
0	1	7				

Key: 30|3 means 33

If your data consists of pairs of values then you should use a **graph**.

Examples of pairs of values could be ...

A **graph** shows data as a series of points plotted on a grid.

Height and foot length.

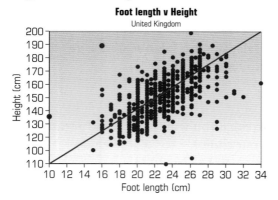
Foot length v Height
United Kingdom

Time of day and height of tide.

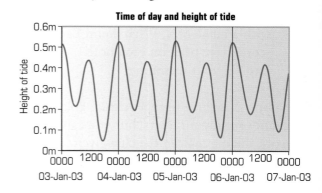
Time of day and height of tide

A **scatter graph** shows up any correlation between two variables. You can draw a **line of best fit** through the points.

In a **line graph** for time series the horizontal axis shows time. You can join the data points with a smooth curve.

Exercise D1.3

1 The table shows the number of animals staying in a vet's surgery one night.

Cats	Dogs	Rabbits	Hamsters	Birds
4	2	1	3	2

Draw a pie chart to represent the data.

2 There are 29 students in Katie's class, and seven of them wear glasses.
Katie wants to show this information on a pie chart.
What angle will Katie need to use for the sector representing students who wear glasses?
Show your working clearly.

3 Ten books were weighed and the number of pages in each was recorded.

Number of pages	660	317	224	222	472	255	419	206	602	437
Weight (g)	453	231	199	230	318	137	426	311	448	333

Draw a scatter graph to represent the data.

4 The table shows the heights of a group of Year 9 students.

Height (cm)	Number of students	Height (cm)	Number of students
$145 \leqslant h < 150$	1	$165 \leqslant h < 170$	23
$150 \leqslant h < 155$	2	$170 \leqslant h < 175$	17
$155 \leqslant h < 160$	18	$175 \leqslant h < 180$	5
$160 \leqslant h < 165$	32	$180 \leqslant h < 185$	2

Represent this data on a frequency diagram.

5 The table shows the temperature of the contents of a test tube during a scientific experiment.

Time (s)	5	10	15	20	25	30	35	40	45	50	55	60	65	70	75	80	85	90
Temp (°C)	18	18	19	24	63	140	189	211	231	244	239	210	177	152	130	112	97	86

Represent this data on a suitable graph.

6 Sketch the type of diagram that you would use for these sets of data.
Explain the reasons for your choice in each case (there may be more than one possibility).
  a A list of the weights (in kg) of the students in a class.
  b A table showing the height (in cm) of a group of students, together with their personal best times for the 100 m sprint.
  c A data collection sheet recording the eye colour for a group of students.
  d A table showing the finishing times for 100 runners in a marathon.
  e A questionnaire recording the numbers of children in the respondents' families.

This spread will show you how to:
▶▶ Interpret graphs and diagrams and draw inferences.
▶▶ Have a basic understanding of correlation.

KEYWORDS
Inference   Exception
Anomaly

▶ An **inference** is a statement based on evidence.

You need to be able to make sense of graphs and charts.

example

These charts show results from a survey about whom adults and children in the UK spend their time with. Karim makes two inferences from the charts:

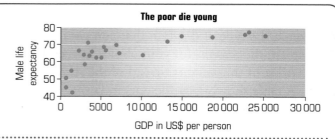

a   For the vast majority of their time, children are in the company of their parents or teachers.

b   At any given time, more adults are alone than children.

These are 'doughnut' charts – pie charts with a hole in the middle!

Comment on the fairness of each of Karim's inferences.

a   This does seem to be a fair inference. Children are likely to be with parents or teachers while they are 'sleeping, working or studying'.

b   This is an unfair inference because the numbers of adults and children are not specified.

Graphs and charts often highlight exceptions to a general trend.

▶ An **anomaly** is a value that does not conform to a rule.

example

The scatter graph shows the male life expectancy and GDP per person (a measure of wealth) for 23 different countries.
Is the graph title fair?

The inference that life expectancies are lower in poorer countries seems to be confirmed by the data. There is an anomaly: countries with a GDP per person between about $2000 and $5000 have quite a range of life expectancies.

Exercise D1.4

1 The chart shows the amount of time that people in the UK spend on different activities in a typical day.

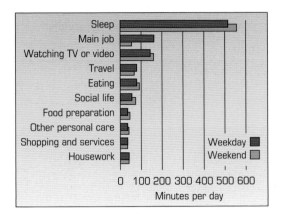

Describe the main differences between the way people use their time during weekdays and weekends.

Suggest reasons for the differences.

2 The graph shows the life expectancies of people born in the UK in the years from 1901.

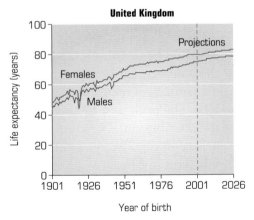

Explain the main features of the graph.

3 The chart shows temperature information for a village in England.

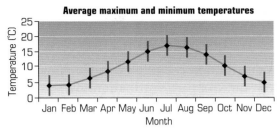

Explain how the graph works, and describe its main features.

4 Sketch a scatter graph for each of these relationships.
   a Televisions with bigger screens cost more money.
   b Older cars cost less than new cars.
   c Cakes take less time to cook if the oven is hotter.
   d The distance you can throw a tennis ball does not depend on your height.
   e Students who do well in a history test usually do well in geography.

5 The graph shows the number of people at a railway station one morning.

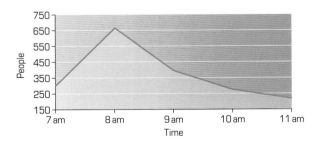

Are these statements true or false? Explain your answers.
   a The number of people in the station more than doubled between 7 am and 8 am.
   b The number of people in the station halved between 8 am and 9 am.

Comparing distributions

This spread will show you how to:

▶▶ Compare two or more distributions and make inferences using the shape of the distributions, the range and appropriate statistics.

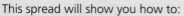

KEYWORDS
Distribution   Modal class
Range
Population pyramid
Compare

A distribution is a graph or table showing data frequencies. You can compare distributions by:

▶ identifying key features in a diagram
▶ calculating statistics.

example

The table shows the population (in thousands) of Brazil and the UK, broken down by age group.

Compare the distribution of population in the two countries.

Age	Brazil	UK
0–14	50 625	11 268
15–29	50 348	11 467
30–44	38 525	13 595
45–59	22 191	11 039
60–74	10 754	7 775
75+	3 110	4 379
Total	**175 553**	**59 523**

The table shows that the population of Brazil is about three times as big as the UK population.

▶ Show the information in a population pyramid:

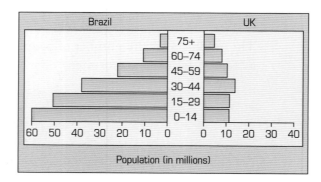

▶ Comment on the key features of the diagram:

Brazil: The distribution peaks at the youngest age groups, and tails off as the ages increase.
UK: The peak is in the middle of the distribution. There are more people in the 75+ age group in the UK than in Brazil.

You cannot find the range since you do not have the raw data.

▶ Use appropriate statistics.

The **modal class** is the class with the highest frequency.

The modal class is the 0–14 age group for Brazil, and the 30–44 age group for the UK.

You could have chosen different diagrams, such as a pair of pie charts.

Exercise D1.5

1 The customer rating of two types of car, from 0 (worst) to 10 (best) is:

Type A	7	8	7	7	9	8	7	8
Type B	2	1	5	8	10	7	4	0

  a Find the average rating for each car.
  b Find the range for each car.
  c Which type of car were customers happier with? Explain your answer.

2 A group of students investigated four different brands of crisps. For each brand, they opened ten packets of crisps, and counted the number of crisps in the packet. The results are shown in the table.

Brand A	19 17 19 17 18 19 18 18 17 18
Brand B	17 16 14 20 19 20 19 17 17 18
Brand C	21 21 21 21 20 23 22 20 20 21
Brand D	15 15 17 15 16 18 17 18 15 21

  a For each brand, find the average number of crisps per packet.
  b Find the range of the numbers of crisps per packet.
  c It would not be sensible to say 'The brand with the highest average number of crisps in a packet represents the best value for money'. Explain why not.

3 A class took two different tests on the same day.
The results are shown in the table.

Test A	70 81 69 72 81 74 63 82 73 88 75 72 70 71 63 77 70 86 68 76 85 73 71 72 78
Test B	50 61 62 68 44 68 59 54 47 54 51 54 58 55 57 75 74 57 57 47 75 47 43 59 36

  a Explain why the mean would be a suitable average for these sets of data.
  b Calculate the mean of each set, using an appropriate assumed mean.
  c Find the range of each set of marks.
  d Explain in words which test you think was more difficult.

4 20 boys and 20 girls took part in a sports tournament in which the aim was to score points. The results are shown in the table.

Number of points	5	6	7	8	9	10	11	12	13	14
Number of boys	1	3	1	2	2	2	2	3	1	3
Number of girls	0	0	4	2	4	3	3	3	1	0

Use the information in the table to compare the performance of the boys and the girls. Comment on the distribution of the scores, and calculate appropriate statistics.

5 There are 30 students in class 9A, and 12 of them own cats.
  a Draw a pie chart to show this data.
  b When a similar chart is drawn for the students in class 9B, the angle for cat owners is 180°. Can you say 'There must be more cat owners in class 9B than there are in class 9A'? Explain your answer.
  c There are 30 students in class 9C. The angle for cat owners on a pie chart for class 9C is 168°. Is it possible to say 'People in class 9C own more cats than people in class 9A'? Explain your answer.

6 The chart shows the heights of a set of adults.
Use the information in the chart to decide whether these statements are true or false, or whether it is impossible to tell. Explain your answers.

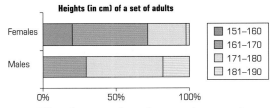

Heights (in cm) of a set of adults

  a The shortest female was shorter than the shortest male.
  b The tallest male was taller than the tallest female.

This spread will show you how to:
▶▶ Communicate interpretations and results of a statistical enquiry using selected tables, graphs and diagrams in support.

KEYWORDS

Axes   Sample
Fair   Raw data
Interpret

Statistics can be used in a misleading way.
The tourist board in Apton produced this advertisement:

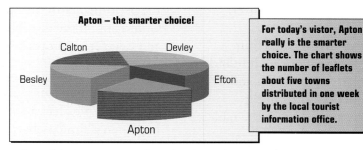

Apton – the smarter choice!

Calton   Devley
Besley   Efton
Apton

For today's vistor, Apton really is the smarter choice. The chart shows the number of leaflets about five towns distributed in one week by the local tourist information office.

To decide whether the advertisement is fair, look at the **raw data**:

Apton had the **smallest** number of leaflet requests.
To create a better impression the tourist board:

▶ Used a 3-D chart, and put Apton at the front.
▶ 'Exploded' the Apton sector, made it a brighter colour, and gave it a bigger label.
▶ Chose a title for the graph to suggest how it should be interpreted.

Town	Leaflets
Apton	24
Besley	28
Calton	25
Devley	26
Efton	26

Altering the axes on a graph can often lead to a distorted picture.
These graphs tell different stories about the British cinema industry.

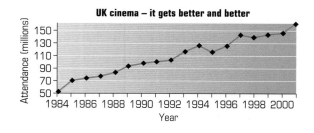

UK cinema – it gets better and better

UK cinema – back from the brink?

The vertical scale on this graph does not start at zero, so it exaggerates the rise.

The horizontal scale on this graph shows a much wider range, giving a fuller picture.

Statistics can also be used in a misleading way.
Here is an advert for a brand of cat food.

75% of cat owners said that their cats had more energy after trying new KattiChunks!

The claim may or may not be fair.
You are not told how the information was gathered, or how big the sample was.

Exercise D1.6

1 Each of these statements is supported by a misleading statistical diagram.
Explain what is wrong with each one, and draw a fairer diagram to represent the data.

a 'Sales of *The Daily Bugle* just get better and better!'

Month	Average daily sale
Jan	500
Feb	550
Mar	650
Apr	680
May	690
Jun	720
Jul	740

b 'Customers love our great service.'

Customer	Ratings
Poor	28%
Fair	52%
Good	8%
Superb	12%

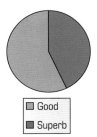

□ Good
■ Superb

c 'All the local beaches got great marks for cleanliness, but Apton's famous beach really leads the way!'

Clean beach scores	
Apton	87%
Besley	91%
Calton	92%
Devley	91%
Efton	95%

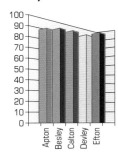

2 The Sandwell Bay Tourist Board records the number of people on the beach and in the pool each day one month in a scatter graph.

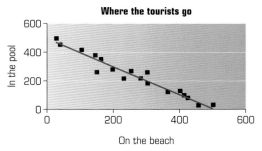

The leisure pool has to close for one day for maintenance. The manager of a beach café says: 'With the pool shut, we will expect about 500 people on the beach.' Explain why the manager may be wrong.

3 For each of these sets of numbers, work out the mean, the median and the mode.

Set 1
1, 1, 2, 2, 2, 4, 5, 7, 9, 48

Set 2
0, 45, 45, 46, 48

Set 3
502, 513, 519, 1208

Set 4
29, 33, 48, 21, 33, 56, 62

Explain which of these averages you would use with each set:
a to give the impression that the numbers were large
b to give the impression that the numbers were small
c to give as fair an impression as possible.

Summary

You should know how to …

1 Design a survey or experiment to capture the necessary data from one or more sources.

2 Determine the sample size and degree of accuracy needed.

3 Design, trial, and if necessary refine data collection sheets.

4 Communicate interpretations and results of a statistical enquiry using selected tables, graphs and diagrams in support.

Check out

1 Tom and Ben decide to test their hypothesis that:

In Year 9, boys watch more TV per day than girls.

Describe what is meant by primary and secondary data, and give examples of how each of these could be used in testing their hypothesis.

2 Tom says, 'We could go to our friends' houses in the holidays and see how much TV they watched!'

Ben says, 'We could send a questionnaire to all the Year 9 students at our ten nearest schools to find out how much TV they watched everyday.'

a Give one advantage and one disadvantage of:

i Tom's method

ii Ben's method.

b Suggest your own method for gathering data. Remember to consider the size and type of sample.

3 a Explain how Tom and Ben could use a pilot survey to help in the design of a questionnaire or data collection sheet.

b Write an example of a survey question that they could use. Make sure it presents a series of options to choose from. Explain why the question would be set out in this way.

4 a Tom and Ben decide to illustrate their project using some 3-D charts produced on a computer spreadsheet. Give one advantage, and one disadvantage, of using 3-D charts. Use diagrams to help explain your answer.

b Explain how a statistical diagram can give a misleading impression, even if the data in the chart are completely accurate.

Perimeter, area and volume

This unit will show you how to:

▶▶ Use units of measurement to calculate, estimate, measure and solve problems in a variety of contexts.

▶▶ Convert between area measures and between volume measures.

▶▶ Know and use the formulae for the circumference and area of a circle.

▶▶ Calculate the surface area and volume of right prisms.

▶▶ Solve increasingly demanding problems and evaluate solutions.

▶▶ Solve substantial problems by breaking them into simpler tasks.

▶▶ Present a concise, reasoned argument using symbols, diagrams and text.

Wheels are circular – they have one continuous edge that can go round very smoothly!

Before you start

You should know how to ...

1 Find the perimeter and area of a shape.

2 Convert between units of length.

3 Find the area of a triangle.
   ▶ Area = $\frac{1}{2}$ × base × height

4 Find the volume of a cuboid.
   ▶ Volume = length × width × height

Check in

1 Find the perimeter and area of this shape.

6 m
2 m
4 m
10 m

2 Change:
   a 120 cm to m   **b** 26.1 km to m
   c 37 mm to cm   **d** 19.2 m to cm

3 Find the area of these triangles.
   a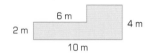
   8 cm
   5 cm
   b
   5 cm
   4 cm

4 Find the volumes of these cuboids.
   a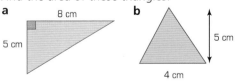
   1 m
   3 m
   5 m
   b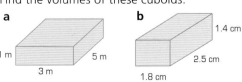
   1.4 cm
   2.5 cm
   1.8 cm

This spread will show you how to:
- ▶▶ Use units of measurement to calculate, estimate, measure and solve problems in a variety of contexts.
- ▶▶ Convert between area measures.

KEYWORDS

Area	Convert
Length	Parallelogram
Trapezium	Hectare

You should know how to find the areas of these shapes:

Triangle

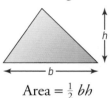

Area $= \frac{1}{2}bh$

Parallelogram

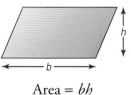

Area $= bh$

Trapezium

Area $= \frac{1}{2}(a+b)\,h$

To use the formulae, all the lengths must be in the same unit.

- ▶ The metric units of length are mm, cm, m and km.
- ▶ The metric units of area are mm², cm², m² and km².

You should know these metric conversions:
- ▶ 1 cm = 10 mm
- ▶ 1 m = 100 cm
- ▶ 1 km = 1000 m

You can convert between metric units of area.

This square is 1 cm².
Each side is 1 cm.

Each side is 10 mm.
The area is 10 mm × 10 mm = 100 mm²

This square is 1 m².
Each side is 1 m.

Each side is 100 cm.
The area is 100 cm × 100 cm = 10 000 cm²

- ▶ 1 cm² = 100 mm²   1 m² = 10 000 cm²

example

Change:
a 25.6 cm² to mm²   **b** 64 000 cm² to m²

....

a Imagine a rectangle:

25.6 cm

1 cm

Change the lengths to mm:

256 mm

10 mm

25.6 cm² = 2560 mm²

b Imagine a rectangle:

640 cm

100 cm

Change the lengths to m:

6.4 m

1 m

64 000 cm² = 6.4 m²

Choose one length to be 100 cm (= 1 m)

Exercise S2.1

1 Calculate the areas in cm^2 of each of these shapes.

a

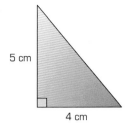

5 cm

4 cm

b

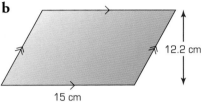

12.2 cm

15 cm

c

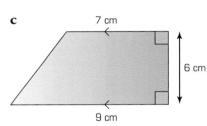

7 cm

6 cm

9 cm

2 Change these areas to mm^2.
  a 6 cm^2 **b** 26 cm^2
  c 0.6 cm^2 **d** 14.6 cm^2
  e 240 cm^2

3 Change these areas to cm^2.
  a 4 m^2 **b** 7.3 m^2
  c 0.56 m^2 **d** 17.2 m^2
  e 256 m^2

4 Change these areas to m^2.
  a 60 000 cm^2 **b** 125 000 cm^2
  c 6000 cm^2 **d** 2 000 000 cm^2
  e 450 cm^2

5 Change these areas to cm^2.
  a 600 mm^2 **b** 7500 mm^2
  c 85 mm^2 **d** 120 000 mm^2
  e 44 mm^2

6 Calculate the area in cm^2 of each shape.

a

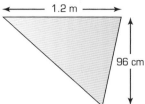

1.2 m

96 cm

b

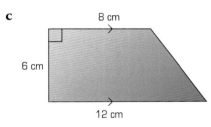

63 cm

3 m

c

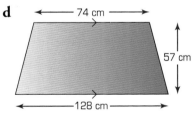

8 cm

6 cm

12 cm

d

74 cm

57 cm

128 cm

  e Change each of your answers to:
    i mm^2 **ii** m^2

7 Use diagrams to show why the formulae for the area of a triangle, parallelogram and trapezium work.

8 **Investigation**
A mathematician   1 hectare = 10 000 m^2
has a field with an area of 1 hectare. Investigate possible dimensions for his field.

Circumference of a circle

This spread will show you how to:

▶▶ Know and use the formula for the circumference of a circle.

KEYWORDS

Circle   Radius
Perimeter   Centre
Diameter   Circumference
Pi (π)

A circle is a set of points which are equidistant from a fixed point, the centre.

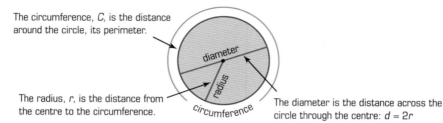

The circumference, C, is the distance around the circle, its perimeter.

diameter

radius

circumference

The radius, r, is the distance from the centre to the circumference.

The diameter is the distance across the circle through the centre: $d = 2r$

The relationship between the circumference and the diameter is a special number.
It is just over 3. Good approximations are 3.14 or $\frac{22}{7}$.

You can use the symbol π (pi) for the exact value:

The exact value of π cannot be written down – its decimal places go on for ever without making a pattern.

▶ Circumference = $\pi \times$ diameter, $C = \pi d$
▶ Circumference = $\pi \times 2 \times$ radius, $C = 2\pi r$

You can use the π button on your calculator, or an estimate of the value of π to find the circumference.

example

Find the circumferences of these circles:

a

12 m

b

3.5 cm

....

a $C = \pi d$
$C = \pi \times 12$ m

Using $\pi = 3$:
$C = 36$ m

Using the calculator button:
$C = 37.70$ m (to 2 dp)

b $C = 2\pi r$
$C = 2 \times \pi \times 3.5$ cm = 7π cm

Using $\pi = \frac{22}{7}$:
$C = 7 \times \frac{22}{7}$ cm = 22 cm

Using the calculator button:
$C = 21.99$ cm (to 2 dp)

Exercise S2.2

1 Write down the radius and diameter of each of these circles.

a
6 cm

b
16 cm

c
9 m

d
27 cm

2 Find the circumference of each of these circles.

a
11 cm

b
7 cm

c
4.2 cm

d
21.2 cm

3 A big wheel has a diameter of 30 m.

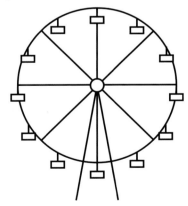

How far do you travel in one revolution of the wheel?

4 The radius of a bicycle wheel is 15 inches. What is the circumference of the wheel?

5 To find the diameter of a circle from its circumference, you rearrange the formula:

$$C = \pi d$$
$$\text{so } \frac{C}{\pi} = d$$

Find the diameter of a circle with circumference:
a 27 cm   b 14 cm   c 27.1 m

6 The largest big wheel in Scarborough has a circumference of 60 m.
Find the radius of the wheel.

7 The Golden Dragon Restaurant has a round table with a diameter of 2.7 m. How many people can sit round the table? Assume each person needs 45 cm of space.

8 Find the perimeter of each of these shapes.

a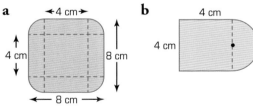
4 cm
4 cm
8 cm
8 cm

b
4 cm
4 cm

c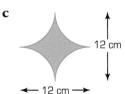
12 cm
12 cm

Perimeter is the distance around the edge of a shape.

9 **Challenge**
The diameter of the wheel of a racing bike is 60 cm.
How many rotations does the wheel make in a 50 km race?

Area of a circle

This spread will show you how to:

▶▶ Know and use the formula for the area of a circle.

KEYWORDS

Circle   Radius
Sector   Diameter
Circumference   Pi (π)

The formula for the area, A, of a circle is:

Area $= \pi \times$ radius $\times$ radius

You write this as:
$A = \pi \times r \times r$ OR πr^2

▶ $A = \pi r^2$

You can also find the area from the diameter.
The diameter is double the radius.

$d = 2r$ so $r = \frac{d}{2}$

$A = \pi \times \left(\frac{d}{2}\right)^2 = \dfrac{\pi d^2}{4}$

▶ $A = \dfrac{\pi d^2}{4}$

The Greeks found this formula over 2000 years ago.
You can split the circle into lots of tiny sectors ...

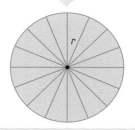

and then rearrange them into a 'rectangle'.

'rectangle'

half circumference $= \pi r$

Area of 'rectangle' $= l \times b$
                   $= \pi r \times r$
                   $= \pi r^2$

example

Find the areas of these circles:

a

b

....

a $A = \pi r^2$
    $r = 6$ cm
    $A = \pi \times 6$ cm $\times 6$ cm
    $A = \pi \times 36$ cm^2

Using $\pi = 3$:
$A = 108$ cm^2

Using π on a calculator:
$A = 113.10$ cm^2 (2 dp)

b $A = \pi r^2$
    $r = 14$ m $\div 2 = 7$ m
    $A = \pi \times 7$ m $\times 7$ m
    $A = \pi \times 49$ m^2

Using $\pi = \frac{22}{7}$:
$A = \frac{22}{7} \times 49$ m^2
   $= 22 \times 7$ m^2
   $= 154$ m^2

Using π on a calculator:
$A = 153.93$ m^2 (2 dp)

Exercise S2.3

1 Find the circumference of each of these circles.

a

8 cm

b

26 mm

c

2.1 m

d

20 m

2 Find the area of each circle in question 1.

3 Use π as 3 to estimate the circumference and the area of each of these circles.

a

10 cm

b

8 cm

c

4 mm

d

20.7 cm

4 Find the shaded areas.

a
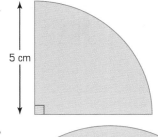
5 cm

b
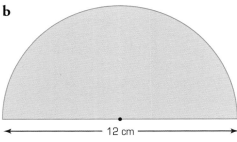
12 cm

c
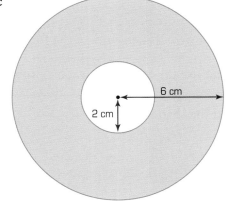
6 cm
2 cm

5 A small pizza has a diameter of 22 cm. A large pizza has a diameter of 32 cm.
  a Find the area of the top of each pizza.
  b The small pizza costs £5. The large pizza costs £9.
     Which is better value and why?
  c Choose a suitable size and price for a medium pizza. Explain your reasoning.

6 **Challenge**
  ▶ What happens to the area of a circle if you double the radius?
  ▶ What happens to the area of a circle if you multiply the radius by 3?
  Explain and justify your answers.

This spread will show you how to:
▸▸ Use units of measurement to calculate, estimate, measure and solve problems in a variety of contexts.
▸▸ Know and use the formulae for the circumference and area of a circle.

KEYWORDS
Circle   Radius
Area   Formulae
Circumference   Pi (π)

You should learn these circle formulae:

▸ $C = \pi d$

▸ $A = \pi r^2$

You should be able to work out that:

▸ $C = 2\pi r$

▸ $A = \dfrac{\pi d^2}{4}$

It is easy to mix up the two formulae.

Hint
Circumference is a length so it only has one dimension, d.
Area has two dimensions so the formula includes r^2.

You can use the formulae to solve problems involving circles.

example

Find the shaded areas.

a

5 cm

b

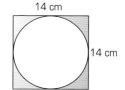

14 cm

14 cm

a You want to find $\frac{3}{4}$ of the area of the whole circle.

Find the whole area:
$A = \pi r^2$
$A = \pi \times 5^2$
$A = 78.5398... \text{ cm}^2$

$\frac{3}{4}$ of the area
$= \dfrac{3 \times 78.5398...}{4} \text{ cm}^2$
$= 58.9084... \text{ cm}^2$
$= 58.9 \text{ cm}^2 \text{ (1 dp)}$

b Shaded area = area of square − area of circle
Area of square = 14 cm × 14 cm
        = 196 cm^2
Area of circle = πr^2
Radius   = 14 cm ÷ 2 = 7 cm
Area of circle = $\pi \times 49 \text{ cm}^2$
        = 153.9 cm^2 (1 dp)
Shaded area   = 196 cm^2 − 153.9 cm^2
        = 42.1 cm^2 (1 dp)

Exercise S2.4

1 The radius of a big wheel is 37 m. Joe starts at the bottom. The big wheel stops when he is at the top. How far has he travelled to reach the top?

2 **a** Find the area of a circle with radius 6 m.
  b Find the area of a semicircle with radius 6 m.

3 The minute hand of this clock is 10 cm long.

  a How far does the tip of the minute hand travel in 1 hour?
  b The hour hand is 4 cm long. How far will the tip of the hour hand travel in
  i 1 hour **ii** a day?

4 Find the shaded areas.

  a

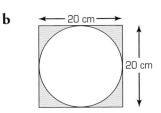

  b

  20 cm

  20 cm

  c

  12 cm

  12 cm

5 A dance record is made from vinyl.

The diameter of the record is 30 cm. The hole in the middle has a diameter of 0.9 cm.
  a Find the circumference of the record.
  b Find the area of the hole.
  c Find the area of the vinyl.

6 **a** Calculate the area of this church door.
  b Calculate the perimeter of the door.

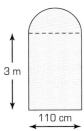

  3 m

  110 cm

7 A dining room table is a rectangle with a semicircle at each end.

  40 cm

  120 cm

Calculate the area of the dining table.

8 You have 160 m of fencing to enclose a field.
  What area will you enclose if:
  a the field is square
  b the field is circular?

9 Repeat question 8 for different lengths of fencing. What do you notice?

This spread will show you how to:
▶▶ Use units of measurement to calculate, estimate, measure and solve problems in a variety of contexts.
▶▶ Convert between volume measures.

KEYWORDS
Length   Area
Volume   Dimensions
Capacity

▶ Length has one dimension.
The metric units of length are mm, cm, m and km.

▶ Area has two dimensions.
The metric units of area are mm², cm², m² and km².

You measure area in squares:

$1 \text{ cm}^2 = 100 \text{ mm}^2$

$1 \text{ m}^2 = 10\ 000 \text{ cm}^2$

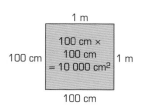

1 m

100 cm × 100 cm = 10 000 cm²

100 cm

1 m

100 cm

▶ Volume has three dimensions.
The metric units of volume are mm³, cm³, m³ and km³.

> You say m³ as 'cubic metres'.

You measure volume in cubes:

This cube is 1 cm³.
Each side is 1 cm or
each side is 10 mm.

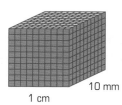

10 mm

1 cm

This cube is 1 m³.
Each side is 1 m or
each side is 100 cm.

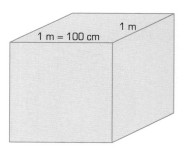

1 m = 100 cm

1 m

The volume is
10 mm × 10 mm × 10 mm = 1000 mm³

The volume is
100 cm × 100 cm × 100 cm = 1 000 000 cm³

example

Change:

a   25.3 cm³ to mm³

b   640 000 cm³ to m³.

....

a   Imagine a cuboid:

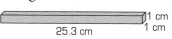

1 cm
25.3 cm
1 cm

25.3 cm × 1 cm × 1 cm
= 253 mm × 10 mm × 10 mm
= 25 300 mm³

b   Imagine a cuboid:

Choose lengths to be 100 cm = 1 m.
100 cm × 100 cm × 64 cm
= 1 m × 1 m × 0.64 m
= 0.64 m³

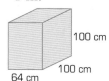

100 cm

100 cm

64 cm

Exercise S2.5

1 Calculate the volume of each of these cuboids in cubic centimetres.

a
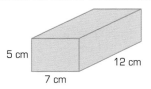
5 cm
12 cm
7 cm

b
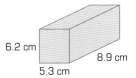
6.2 cm
8.9 cm
5.3 cm

c
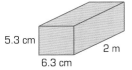
5.3 cm
2 m
6.3 cm

2 Give your answers to question 1 in cubic millimetres.

3 Calculate the volume of each of these cuboids in cubic centimetres.

a
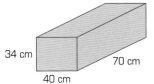
34 cm
70 cm
40 cm

b
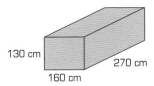
130 cm
270 cm
160 cm

c
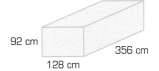
92 cm
356 cm
128 cm

4 Give your answers to question 3 in cubic metres.

5 Change these volumes to cubic millimetres.
  a 6 cm^3 **b** 26 cm^3
  c 0.7 cm^3 **d** 14.6 cm^3
  e 250 cm^3

6 Change these volumes to cubic centimetres.
  a 3 m^3 **b** 6.5 m^3
  c 0.63 m^3 **d** 15.6 m^3
  e 46 m^3

7 Change these volumes to cubic centimetres.
  a 6000 mm^3 **b** $25\,000 \text{ mm}^3$
  c 700 mm^3 **d** $56\,000 \text{ mm}^3$
  e 75 mm^3

8 Change these volumes to cubic metres.
  a $6\,000\,000 \text{ cm}^3$ **b** $17\,600\,000 \text{ cm}^3$
  c $450\,000 \text{ cm}^3$ **d** 450 cm^3
  e 70 cm^3

9 **Investigation**

> $1 \text{ litre} = 1000 \text{ cm}^3$

A container in the shape of a cuboid has a capacity of 1 litre.

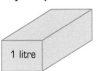
1 litre

1 litre

Investigate possible dimensions for the container.

10 **Challenge**
A can in the shape of a cylinder has a capacity of 1 litre. Investigate possible dimensions for the container.

FANTA-STIC

> Area of a circle = πr^2.

This spread will show you how to:
▶▶ Calculate the surface areas and volumes of right prisms.

KEYWORDS
Prism   Face
Surface area   Cross-section

A prism is a 3-D shape with a constant cross-section throughout its length.

You should know these common prisms:

You can name a prism by its cross-section.

cuboid   triangular prism   hexagonal prism   cylinder

For a prism:
▶ **Volume = area of cross-section × length**
▶ **Surface area = total area of all the faces**

A cuboid is a rectangular prism.
Volume = length × width × height

The end face of a prism is its cross-section.

Find the volume and surface area of these shapes:

a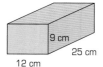

b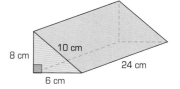

....

a Volume = length × width × height
= 12 cm × 9 cm × 25 cm
= 2700 cm^3

There are three different faces:

Surface area
= 2 × 108 cm^2 + 2 × 225 cm^2
+ 2 × 300 cm^2
= 1266 cm^2

b Volume = area of cross-section × length
Cross-section is a triangle.
Area   = $\frac{1}{2}$ × 6 cm × 8 cm = 24 cm^2
Volume = 24 cm^2 × 24 cm
= 576 cm^3

There are four different faces:

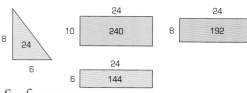

Surface area
= 2 × 24 cm^2 + 240 cm^2 + 144 cm^2
+ 192 cm^2
= 624 cm^2

Exercise S2.6

1 Work out the volumes of these cuboids.

a

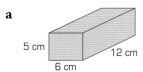
5 cm
12 cm
6 cm

b

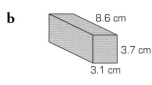
8.6 cm
3.7 cm
3.1 cm

c

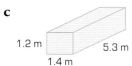
1.2 m
5.3 m
1.4 m

2 Calculate the surface areas of the cuboids in question 1.

3 Work out the volumes of these prisms.

a

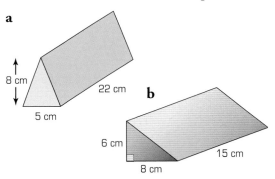
8 cm
22 cm
5 cm

b
6 cm
15 cm
8 cm

c

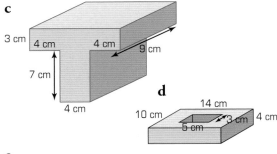
3 cm
4 cm
4 cm
9 cm
7 cm
4 cm

d
14 cm
10 cm
3 cm
4 cm
5 cm

e

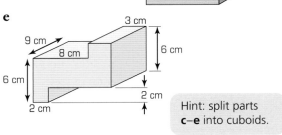
3 cm
9 cm
8 cm
6 cm
6 cm
2 cm
2 cm

Hint: split parts **c–e** into cuboids.

4 Find the surface areas of the prisms in question 3 parts **c** and **e**.

5 Find the lengths marked with letters.

a

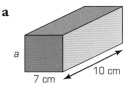
a
7 cm
10 cm

volume = 420 cm^3

b

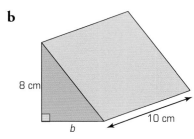
8 cm
10 cm
b

volume = 240 cm^3

c

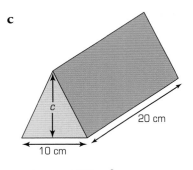
c
20 cm
10 cm

volume = 1600 cm^3

6 Investigation
a A cuboid has a volume of 600 cm^3. Investigate its possible dimensions.
b Change the volume and investigate possible dimensions.

7 Investigation
a A prism has a volume of 600 cm^3. Investigate its possible dimensions.
b Change the volume and investigate possible dimensions.

You should know how to ...

1 Know and use the formulae for the circumference and area of a circle.
▶ $C = \pi d$
▶ $A = \pi r^2$

2 Solve substantial problems by breaking them into simpler tasks.

3 Present a concise, reasoned argument using symbols, diagrams and related explanatory text.

Check out

1 At the Zing Vaa restaurant each diner needs at least 70 cm of table.

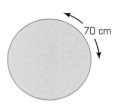

a Calculate the number of people that can be seated around a circular table of diameter 3 m.

b How many people can be seated around this table?

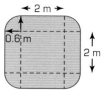

c How many people can be seated around this table?

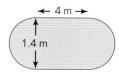

2 ABCD is a rhombus.

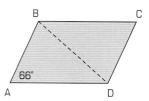

Find angle BDC.

3 Do you think 50 people could sit around a circular table with a diameter of 6 m? Explain and justify your answer.

Number calculations

This unit will show you how to:

- Extend knowledge of integer powers of 10.
- Multiply and divide numbers by any integer power of 10.
- Use rounding to make estimates.
- Round numbers.
- Use ICT to estimate square roots.
- Use index notation for integer powers and simple instances of the index laws.
- Know that a recurring decimal is an exact fraction.
- Extend mental methods of calculation.
- Use known facts to derive unknown facts.
- Make and justify estimates and approximations of calculations.

- Use standard column procedures to add and subtract integers and decimals of any size.
- Multiply and divide by decimals, dividing by transforming to division by an integer.
- Use a calculator efficiently and appropriately.
- Use a range of function keys on the calculator.
- Enter numbers and interpret the display.
- Solve word problems mentally.
- Check results using appropriate methods.
- Give solutions to problems to an appropriate degree of accuracy.
- Use trial and improvement methods where a more efficient method is not obvious.

How much is that then?

Say £800 + VAT.
10% is 80, 5% is £40, 2.5% is £20.
The VAT is £140 so it's under a grand.

Estimating and mental methods are useful everyday skills.

Before you start

You should know how to ...

1 Add and subtract negative numbers.
   - Adding a negative = subtracting a positive
   - Subtracting a negative = adding a positive

2 Find equivalent fractions.

$$\frac{2}{3} \xrightarrow[\times 6]{\times 6} = \frac{12}{18}$$

3 Use standard procedures to add, subtract, multiply and divide integers.

Check in

1 Work out:
   a $^-2 + {}^-5$ **b** $2 - 5$ **c** $^-3 - 7$
   d $3 - {}^-7$ **e** $-\frac{1}{5} + \frac{2}{5} - {}^-\frac{4}{5}$

2 Write these fractions with a common denominator of 36:
   $\frac{1}{2}$ $\frac{2}{3}$ $\frac{3}{4}$ $\frac{5}{6}$
   Which is the largest fraction?

3 Calculate without a calculator:
   a $235 + 879$ **b** $1956 - 237$
   c 234×41 **d** $253 \div 11$

This spread will show you how to:

▶▶ Use index notation for integer powers and simple instances of the index laws.

▶▶ Use a calculator efficiently and appropriately.

▶▶ Use trial and improvement.

KEYWORDS

Cube number Power

Indices Index form

Square root Cube root

Index notation

You use index notation to represent powers of a number.

Positive powers

$17^4 = 17 \times 17 \times 17 \times 17 = 83\,521$

$(^-3)^5 = {}^-3 \times {}^-3 \times {}^-3 \times {}^-3 \times {}^-3 = {}^-243$

Negative powers

$10^{-2} = \frac{1}{10^2} = \frac{1}{10 \times 10} = \frac{1}{100} = 0.01$

▶ Any number raised to the power of zero is 1:

$10^0 = 1 \qquad 4^0 = 1$

You can multiply and divide numbers written in index form.

Add the indices when multiplying:

$5^3 \times 5^4 = 5^{3+4} = 5^7$

$5^3 \times 5^4 = (5 \times 5 \times 5) \times (5 \times 5 \times 5 \times 5)$
$= 5 \times 5 \times 5 \times 5 \times 5 \times 5 \times 5$
$= 5^7$

Subtract the indices when dividing:

$4^2 \div 4^5 = 4^{2-5} = 4^{-3}$

$4^2 \div 4^5 = \frac{(4 \times 4)}{(4 \times 4 \times 4 \times 4 \times 4)}$
$= \frac{(^1\!4 \times {}^1\!4)}{(4 \times 4 \times 4_1 \times 4_1 \times 4)}$
$= \frac{1}{4^3} = 4^{-3}$

You can input powers on a calculator using the $\boxed{x^y}$ key:

$17^4 = 17\ \boxed{x^y}\ 4$ 　　　 $(^-3)^5 = 3\ \boxed{+/-}\ \boxed{x^y}\ 5$ 　　　 $10^{-2} = 10\ \boxed{x^y}\ 2\ \boxed{+/-}$

▶ A positive integer has two square roots, one positive and one negative.

$\sqrt{144} = 12 \quad \text{or} \quad {}^-12$

▶ A positive integer has a positive cube root and a negative integer has a negative cube root.

$^3\sqrt{27} = 3 \quad \text{and} \quad {}^3\sqrt{{}^-27} = {}^-3$

You can use trial and improvement to estimate a square root.

example

Estimate $\sqrt{30}$ to 1 decimal place.

$5^2 = 25$ 　　　　　 so $5 < x$

$6^2 = 36$ 　　　　　 so $6 > x$

$5.5^2 = 30.25$ 　　　 so $5.5 > x$

$5.4^2 = 29.16$ 　　　 so $5.4 < x$

$5.45^2 = 29.7025$ 　 so $5.45 < x$

so $\sqrt{30} = 5.5$ (1 dp)

So $5.45 < x < 5.5$

Exercise N2.1

1 Calculate:
  a 5^2 **b** 3^4
  c 2^8 **d** $(^-3)^4$
  e 0.1^2 **f** $\sqrt{121}$
  g $^3\sqrt{27}$ **h** 17^2
  i 25^2 **j** $15^2 + 16^2$

2 Use prime factor decomposition to:
  a Write each of these numbers using index notation, e.g. $25 = 5^2$
    i 196   **ii** 1728
  b Find the square roots of:
    i 576   **ii** 1296

3 **Puzzle**
Find two consecutive whole numbers whose squares differ by 49.

4 Calculate, using your calculator where necessary:
  a 13^3 **b** 5^{-2}
  c 10^{-1} **d** 17^0
  e 15^4 **f** $^3\sqrt{4913}$
  g $^3\sqrt{^-3375}$

5 Calculate, leaving your answer in index form where appropriate:
  a $3^2 \times 3^4$ **b** $4^3 \times 4^2$
  c $5^6 \div 5^4$ **d** $7^6 \div 7^3$
  e $3^2 + 3^3$ **f** $6^4 \times 6^5$
  g $8^3 - 8^2$ **h** $13^8 \div 13^5$
  i $4^3 \times 3^2$ **j** $4^4 \times 4^2 \div 4^8$

6 **a** Use a trial and improvement method to estimate to 2 decimal places:
    i $\sqrt{20}$
    ii $^3\sqrt{90}$
    iii $\sqrt{1000}$
  b Use the square root and cube root keys on your calculator to check your answers.

7 **Puzzle**
Axel says that 'Adding two numbers and then finding the square root is always the same as finding the square root of each number and then adding'.

Morrisey says that 'Adding two numbers and then finding the square root is never the same as finding the square root of each number and then adding'.

Who do you think is correct?
Explain and justify your answer.

8 **a** Write the value of x and y if
    $256 = 16^2 = 4^x = 2^y$
  b Copy and complete:
    $2^{11} = 2048$
    $2^{12} =$

9 **Investigation**
A cube is made from 1 cm × 1 cm × 1 cm cubes.

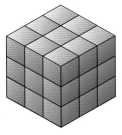

  a The outside of a 3 × 3 × 3 cube is painted red.
How many of the smaller cubes have 3 faces painted? How many of the smaller cubes have 2 faces ... 1 face ... 0 faces painted?
  b Investigate a 4 × 4 × 4 cube.
  c Predict the number of cubes with 2 faces painted in a 5 × 5 × 5 cube ... in a 52 × 52 × 52 cube ...
Explain and justify your prediction.

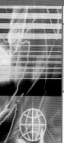

This spread will show you how to:
▶▶ Extend knowledge of integer powers of 10.
▶▶ Multiply and divide numbers by any integer power of 10.

KEYWORDS
Power   Digit
Index   Tenth

You can use index notation to describe smaller powers
of 10 such as tenths and hundredths.

The decimal system is based on powers of 10.

1 hundred	$= 100$	$= 10 \times 10$	$= 10^2$
1 ten	$= 10$	$= 10$	$= 10^1$
1 unit	$= 1$	$= 1$	$= 10^0$
1 tenth	$= \frac{1}{10}$	$= \frac{1}{10}$	$= \frac{1}{10^1}$ $= 10^{-1}$
1 hundredth	$= \frac{1}{100}$	$= \frac{1}{10 \times 10}$	$= \frac{1}{10^2}$ $= 10^{-2}$

These are the values of each of the digits in 4572.836:

10^3 Thousands	10^2 Hundreds	10^1 Tens	10^0 Units	10^{-1} Tenths	10^{-2} Hundredths	10^{-3} Thousandths
4	5	7	2	8	3	6

$\times 10$ (between each column)

The digit 3 stands for 3 hundredths $= 3 \times \frac{1}{100} = 3 \times 10^{-2}$

You can multiply by any integer power of 10.

▶ Multiplying by 0.001 $=$ Multiplying by $\frac{1}{1000}$ $=$ Dividing by 1000
   3.7×0.001 $= 3.7 \times \frac{1}{1000}$ $= 3.7 \div 1000$ $= 0.0037$
▶ Dividing by 0.001 $=$ Dividing by $\frac{1}{1000}$ $=$ Multiplying by 1000
   $0.15 \div 0.001$ $= 0.15 \div \frac{1}{1000}$ $= 0.15 \times 1000$ $= 150$

example

a Calculate $3.702 \div 0.01$.
b Convert $24\,800$ cm^2 into m^2.

....

a $3.702 \div 0.01$ $= 3.702 \div \frac{1}{100}$
      $= 3.702 \times 100$
      $= 370.2$

b $10\,000$ cm^2 $= 1$ m^2
   $24\,800$ cm^2 $= 24\,800 \div 10\,000$ m^2
      $= 2.48$ m^2

Think:
There will be fewer m^2 so divide.

Exercise N2.2

1 Calculate:
  a $46 \div 10$ b 3.8×100
  c $29.7 \div \frac{1}{10}$ d $0.16 \div 10$
  e $13.02 \times \frac{1}{10}$ f 0.27×0.1
  g $1.34 \div \frac{1}{100}$

2 Write down the value of these numbers:
  a 5×10^2 b 2.7×10^3
  c 7.62×10^2 d 1.04×10^5

3 Calculate:
  a 1.7×0.01 b 0.001×43
  c $10\ 000 \times 1.2$ d $8 \div \frac{1}{1000}$
  e 9×0.001 f $23 \times \frac{1}{1000}$
  g $3.7 \div 0.001$ h $0.025 \div 0.01$
  i 0.07×0.01

4 The table shows metric conversions.

1 cm = 10 mm	1 cm^2 = 100 mm^2
1 m = 100 cm	1 m^2 = 10 000 cm^2
1 km = 1000 m	1 km^2 = 1 000 000 m^2

Use the information to convert each of these measurements:
  a 0.32 km into mm
  b 2450 cm^2 into m^2
  c $23\ 500$ mm into km
  d 0.00045 km^2 into cm^2

5 Write down the values of these numbers:
  a 34×10^{-2}
  b 124×10^{-3}
  c 267×10^{-5}
  d 0.024×10^3
  e 6.243×10^2
  f 8.4×10^{-3}

6 Nina is looking at microscopic organisms. The length of a micropod is 0.000517 cm and the length of a didipod is 0.001034 cm.
Use a calculator to work out the difference in length of the two organisms. Give your answer in millimetres.

7 **Puzzle**
Place these numbers in order from lowest to highest to spell something you might need to look at again.
  V $\quad 37.2 \times 10^{-4}$
  R $\quad 2.3 \times 10^{-3}$
  O $\quad 0.0023 \div 10^{-2}$
  I $\quad 4 \times 0.001$
  E $\quad 300 \div 10^5$
  N $\quad 0.0016 \div 0.001$
  I $\quad 110 \div 10^4$
  S $\quad 10.2 \times 10^{-3}$

8 Find the number that lies exactly halfway between each of these pairs of numbers:
  a 24×10^{-1} and 0.037×10^2
  b 3500×10^{-4} and 0.00069×10^3
  c $^-0.34 \times 10^2$ and 1400×10^{-2}
  d $^-274 \times 10^{-2}$ and 0.000115×10^4

9 Copy and complete these statements, giving your answers where appropriate as powers of 10:
  a $0.3 \times ? = 30$
  b $0.4 \times ? = 0.04$
  c $? \div 10^{-2} = 64$
  d $4.3 \times ? = 0.043$
  e $^-2.3 \div ? = ^-2300$
  f $^-0.013 \div ? = 1.3$

This spread will show you how to:
  Use rounding to make estimates.
  Round numbers to the nearest whole number, 1 dp, 2 dp or 3 dp.

KEYWORDS

Round   Digit
Decimal place   Estimate
Nearest   Thousandth
Greater than or equal to ⩾

▶ You can round numbers to a given power of 10.

example

Round 1.27873 to three decimal places.

To round to the nearest thousandth (or 10^{-3}), look at the 10^{-4} digit.
If the 10^{-4} digit is ⩾ 5 round up.
1.27873 is nearer to 1.279 than 1.278
1.27873 ≃ 1.279 (to 3 dp)

```
        1.27873
           ↓
|————————————————————|——+——+——|
1.27              1.278 | 1.279 1.28
                      1.2785
```

You can use rounding to make an estimate.

example

a The bar chart shows the monthly rainfall in Rainsalot. Estimate the total yearly rainfall.

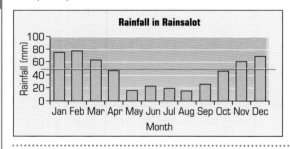

b The weight of a doughnut is 35.6 g. Doughnuts are packed in bags of 6. Bags are packed in boxes that hold 18 bags. Boxes are stacked on pallets that hold 48 boxes. A lorry can hold 11 pallets. Estimate the weight of a lorry-load of doughnuts.

a Average rainfall ≃ 50 mm
Yearly rainfall in Rainsalot
≃ 12 × 50 mm = 600 mm

b Weight of doughnuts in a lorry
≃   40 g × 5 × 20 × 50 × 10
= 2 000 000 grams
= 2000 kg = 2 tonnes

▶ Rounding at the end of a calculation is more accurate.

Rounding during the calculation: $\left(\frac{8}{3}\right)^2$
$\frac{8}{3} = 2.6666...$ = 2.7 (1 dp)
$2.7^2 = 7.29$ = 7.3 (1 dp)

Rounding at the end of the calculation: $\left(\frac{8}{3}\right)^2$
$\frac{8}{3} = 2.6666...$
$(2.6666...)^2 = 7.1111... = 7.1$ (1 dp)

Exercise N2.3

1 Round each of these numbers:
   a to the nearest ten
   b to the nearest whole number
   c to one decimal place
   d to two decimal places.
      i 43.181   ii 9.951
      iii 129.333   iv 12.0999

2 Use your calculator to work out these answers correct to 3 dp:
   a 13% of 2.4465 kg   b $\frac{5}{7}$ of 19 m
   c The weight of a biscuit if 14 biscuits weigh 250 g.

3 a Write down an approximate answer for each of these questions, clearly showing your method of approximation:
      i $\dfrac{24.51 \times 7.8}{2.1}$
      ii $2.7 + (115.3 - 12.78 \times 6.4)$
      iii $\{4.4^2 + (7.2 - 0.738)\}^2$
      iv $\dfrac{4.8 \times 3.7^2}{1.7}$
   b Use your calculator to work out the exact answers, where appropriate giving your answer to 3 dp.

4 A micropod has a length of 0.000517 cm. How many whole micropods would stretch across a line 3 mm long? Use your calculator.

5 a Identify the best estimate to the following calculations from the numbers in the box:

6	7	8		
	9	10	60	

      i $63.23 \div 6.94$
      ii 16.3×0.48
   b Estimate the answer to $\frac{6.56 + 23.7}{3.14}$
   Give your answer to the nearest whole number.

6 **Investigation**
   a When a toilet flushes it uses 9.24 litres of water.
   If a toilet is flushed 17 times a day, about how much water is used per day? Approximately how much water is used per week ... per month ... per year ... per lifetime of a person?
   b The volume of water in Lake Windermere is about 330 000 000 m^3. Approximately how long would it take to flush the volume of the lake down the toilet?

   > $1 \text{ m}^3 = 1000$ litres

7 a The Blakeburn shopping centre has a length of 700 m to the nearest 100 m, and a length of 650 m to the nearest 10 m. Write down a possible length of the Blakeburn shopping centre.
   b Two more shopping centres both have a length of 850 m to the nearest 10 m but have different lengths to the nearest 100 m. Write down a possible length of each of the shopping centres.

8 This diagram shows the football pitch of Greenburgh Utd:

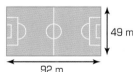

Alf Overestimate, the head groundsman, calculates the area of the pitch as 4579 m^2 to the nearest m^2.
Fred Underestimate, the assistant groundsman, calculates the area of the pitch as 4438 m^2 to the nearest m^2.
Explain how both groundsmen have arrived at their answers.

This spread will show you how to:
- ▶▶ Solve word problems mentally.
- ▶▶ Use standard column procedures to add and subtract integers and decimals of any size.

KEYWORDS

| Partitioning | Integer |
| Compensation | Fraction |

You can add and subtract integers, fractions and decimals mentally.

You should always try to calculate mentally.
Here are some strategies for adding and subtracting mentally:

Partitioning

Breaking a number into parts.
$$^-12.6 - {}^-5.3 = {}^-12.6 + 5.3$$
$$= {}^-12.6 + 5 + 0.3$$
$$= {}^-7.3$$

Compensation

Rounding a number then compensating.
$$0.037 + {}^-0.019 = 0.037 - 0.019$$
$$= 0.037 - 0.02 + 0.001$$
$$= 0.017 + 0.001$$
$$= 0.018$$

Finding a difference by counting up

Counting up from the smaller number to the larger number.

example

Sarah wins £3273 on the lottery and Sam wins £1892.
How much more money does Sarah win than Sam?

$$£3273 - £1892$$
$$= (8 + 100 + 1000 + 273)$$
$$= £1381$$

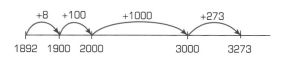

You can use a standard column method for more difficult calculations.

You need a common denominator to add fractions.

example

a Calculate 4829.893 km − 58.6 km − 0.81 km.   **b** Find the total of $\frac{4}{9}$, $^-\frac{2}{3}$ and $\frac{5}{6}$.

a Approximate: $4829.893 - 58.6 - 0.81$
$$\approx 4830 - 59 - 1 = 4770$$

$$^-58.6 - 0.81 = {}^-(58.6 + 0.81) = {}^-59.41$$

$$\begin{array}{r} 4829.893 \\ -59.410 \\ \hline 4770.483 \end{array}$$

4829.893 km − 58.3 km − 0.81 km
= 4770.483 km

b $\frac{4}{9} + \frac{-2}{3} + \frac{5}{6}$
$$= \frac{4}{9} - \frac{2}{3} + \frac{5}{6}$$
$$= \frac{8}{18} - \frac{12}{18} + \frac{15}{18}$$
$$= \frac{8 - 12 + 15}{18}$$
$$= \frac{11}{18}$$

Exercise N2.4

1 Here are five questions you should be able to do in your head. Give yourself about 10 seconds for each question.

   a Find the sum of 1.4, 3.6, 2.6 and 4.4.

   b Write $\frac{28}{35}$ in its simplest form.

   c Vicky has £140. She buys a coat costing £69. How much money does she have left?

   d What is $3000 - 67$?

   e Write three odd numbers that add up to 33.

2 Look at these number cards:

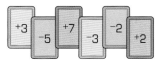

   a What is the total of all six cards?

   b Which three numbers have the lowest possible total?

   c Copy and complete this number sentence using one of the cards to make the lowest possible answer.

3 Calculate each of these, using a mental or written method as appropriate:

   a Two of the angles of a triangle are 46 degrees and 57 degrees. Calculate the size of the missing angle.

   b Three years ago Ralph was 1.36 m tall. Now he is 1.7 m tall.
How much has Ralph grown?

   c Richie earns £342.60 a week. Chloe earns £401.87 a week.
How much more money does Chloe earn each week than Richie?

   d Steve wins £4563 on the lottery. Sue wins £2791. How much more money does Steve win than Sue?

4 In a newspaper, there are three photographs on the same page.

   ▶ Photograph 1 uses $\frac{1}{3}$ of the page.

   ▶ Photograph 2 uses $\frac{1}{6}$ of the page.

   ▶ Photograph 3 uses $\frac{1}{8}$ of the page.

What fraction of the page is not covered by photographs?

5 Calculate these using a mental or written method:

   a $12.17 + 3 + 1.04$

   b $16.4 - 2.53 + {}^{-}1.8$

   c $\frac{2}{7} + \frac{3}{5}$ **d** $2\frac{2}{3} - 1\frac{2}{5}$

   e $24.7 - {}^{-}92.6 + 53.53$

   f $5.79 + 3.7 + 9 + 2.79$

   g $1.8 - 3.17 - 0.93$

   h $1.0562 + 24.92 + 6378$

   i $4567.98 - 67.4 - 0.7$

   j $5.18 \text{ km} + {}^{-}0.0674 \text{ km}$

   k $8305 \text{ m} - {}^{-}340.003 \text{ m} + 0.0032 \text{ m}$

   l $\frac{2}{5} + \frac{3}{4} - \frac{1}{3}$ **m** $1\frac{2}{3} + \frac{-5}{6} - \frac{1}{4}$

6 **Puzzle**

Two families buy tickets for the circus.
The Greens family buy 2 adult tickets and 2 child tickets.
They pay a total of £8.90.
The Browns family buy 1 adult ticket and 3 child tickets.
They pay a total of £8.35.
What is the cost of a child ticket?

7 **Puzzle**

Calculate the perimeter of this shape.
Give your answer in metres.

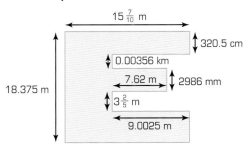

This spread will show you how to:
▶▶ Solve word problems mentally.
▶▶ Extend mental methods of calculation.
▶▶ Use known facts to derive unknown facts.

You should always try to do a calculation mentally.
Here are some strategies for multiplying and dividing:

Factors

Break a number into factors.
$3 \div 0.02$

$0.02 = 2 \times 0.01$
$3 \div 2 = 1.5$
$1.5 \div 0.01 = 150$

Partitioning

Break a number into parts.
21×0.73 m

$(20 \times 0.73$ m$) + 0.73$ m
$= 14.6$ m $+ 0.73$ m
$= 15.33$ m

Near 10s

Round to the nearest 10
then compensate.
$29 \times {}^-0.4$

$30 \times {}^-0.4 - {}^-0.4$
$= {}^-12 - {}^-0.4$
$= {}^-12 + 0.4 = {}^-11.6$

Simplifying fractions during calculation

You can cancel common factors.

example

Jeremy invests £100. His money increases in value by $\frac{7}{8}$.
What is the new value of his money?

It is often easier to calculate with fractions than with decimals as they use whole numbers.

Jeremy's money is now worth $1\frac{7}{8}$ of its original value
$$\frac{15}{8} \text{ of } £100 = \frac{15}{\cancel{8}^2} \times \cancel{100}^{25}$$
$$= \frac{15 \times 25}{2}$$
$$= £187.50$$

Using equivalent calculations

You can change between fractions, decimals and percentages to simplify a calculation.

example

a Calculate the volume of a box that is 2.5 cm by 3.5 cm by 14 cm.

b Calculate 130% of 450 g.

a Volume $= 2.5$ cm $\times 3.5$ cm $\times 14$ cm
$= \frac{5}{2} \times \frac{7}{2} \times 14$ cm^3
$= \frac{5 \times 7 \times \cancel{14}^7}{2 \times \cancel{2}_1}$ cm^3
$= \frac{245}{2}$ cm^3
$= 122.5$ cm^3

b 130% of 450 g $= \frac{130}{100} \times 450$ g
$= 1.3 \times 450$ g
$= 1.3 \times 10 \times 45$ g
$= 13 \times 45$ g
$= 585$ g

Exercise N2.5

1 Here are seven questions you should be able to do in your head. Give yourself about 10 seconds for each question.
  a What is 400 divided by 10?
  b If $3k = 21$, what is the value of $7k$?
  c What is $1 \div 0.02$?
  d If $17 \times 380 = 6460$, how many 17s are there in 3230?
  e Write an approximate answer to the calculation $297.8 \div 4.98$.
  f What is $4^4 \div 16$?
  g Calculate the mean of these numbers: 0, 1, 3, 3, 4.

2 Calculate:
  a 19×11 **b** 0.9×6
  c $^-3 \times 15$ **d** $126 \div 8$
  e 5×1.7 **f** 23×29
  g $35 \div 0.1$ **h** 0.7×30
  i $^-47 \times 0.01$ **j** $224 \div 7$
  k $\frac{3}{7}$ of 105   **l** $232 \times \frac{5}{8}$
  m $^-21 \times 0.7$ **n** $^-42 \div 0.06$
  o $\frac{3}{5} \times ^-24$ **p** $^-8 \div ^-0.04$
  q $^-0.18 \div 0.3$ **r** $^-13 \times 4.1$
  s 5.4×1.5

3 **a** A shop sells flowers at £1.95 a bunch. Find the cost of 25 bunches of flowers.
  b A rectangle is 19 cm long and 11 cm wide.
    Calculate the area of the rectangle.

4 Here are six number cards:

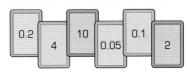

  a Which two cards when multiplied together give the lowest answer?
  b Which two cards, when one is divided by the other, can be used to make a quotient of 200?

5 Calculate:
  a The total weight of 19 pies each weighing 0.4 kg.
  b The total distance covered in 31 turns of a wheel if the circumference of the wheel is 0.92 m.
  c The value of $39x$ if $x = 1.8$.
  d The value of y when $x = 2.7$ if $y = 11x - 5$.
  e Parking costs 15p for every 20-minute period. How much will it cost to park a car from 8.15 am until 5.35 pm?
  f The entrance fee for a cinema is £3.45 per person.
    What is the total cost for a party of 19 people to visit the cinema?
  g Find the number of loaves of bread each weighing $\frac{4}{5}$ kg that can be made using 8 kg of bread mixture.

6 Calculate:
  a $\frac{2}{5}$ of 20.5   **b** 20% of £3.65
  c Increase 350 kg by 30%
  d Decrease 220m by 19%
  e $\frac{7}{20}$ of £375   **f** $\frac{2}{9} \times 189$
  g $\frac{11}{7} \times \frac{23}{19}$ **h** $\frac{12}{25} \div \frac{20}{21}$

7 **Puzzle**
Ben works out 39.7×0.083 on his calculator. His calculator shows an answer of 0.032951.
Explain why this cannot be the right answer.

8 **Puzzle**
Find two numbers whose:
  a sum is 0.7 and whose product is 0.12
  b sum is $^-11.5$ and whose product is 28
  c difference is 28 and whose quotient is 8
  d difference is 6 and whose quotient is $^-0.5$.

This spread will show you how to:
▶▶ Use rounding to make estimates.
▶▶ Multiply decimals.
▶▶ Check results using appropriate methods.

KEYWORDS
Approximate   Equivalent
Estimate   Rounding
Approximately equal to (≈)

You can use equivalent calculations to multiply decimals.
You should always approximate the answer first.

example

Calculate 36.2×0.56

Approximate: $36.2 \times 0.56 \approx 40 \times 0.5 = 20$

Write an equivalent calculation:
$$36.2 \times 0.56 = \frac{362 \times 56}{1000}$$

Use the standard method:

$$
\begin{array}{r}
362 \\
\times 56 \\
\hline
\end{array}
$$

362×50 18100
362×6 2172
$$\underline{20272}$$

The sign ≈ means
'approximately equal to'.

$36.2 = 362 \div 10 = \frac{362}{10}$
$0.56 = 56 \div 100 = \frac{56}{100}$
$\frac{362}{10} \times \frac{56}{100} = \frac{362 \times 56}{10 \times 100}$

Answer: $36.2 \times 0.56 = 362 \times 56 \div 1000$
$= 20\,272 \div 1000$
$= 20.272$

▶ A good estimate will help you judge whether the answer is sensible.

example

a Estimate the value of:
$4.85 \times (5.4 - 0.49)$

b Estimate the area of this trapezium:

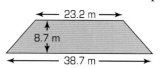

a Approximate:
$4.85 \times (5.4 - 0.49) \approx 5 \times (5) = 25$

b Area of trapezium
$= \frac{1}{2} \times (23.2 + 38.7) \times 8.7$

Approximate area
$= \frac{1}{2} \times (20 + 40) \times 9$
$= \frac{1}{2} \times 60 \times 9$
$= 270 \text{ m}^2$

Exercise N2.6

1 a One tin of dog food weighs 830 g. Karl and Vicky place 59 tins into a box. What is the total weight of the 59 tins in kilograms?

b What is the volume of a box measuring 12 cm in height, 7 cm in length and 5 cm in width?

2 Puzzle

A number is a multiple of 22 and 37, and it has four digits. What is the largest number it could be?

3 Calculate:

a	7×0.01	**b**	6×0.8
c	19×7	**d**	13×15
e	23×17	**f**	1.7×23
g	0.17×23	**h**	29×28
i	6.25×80	**j**	2.3×39
k	3.5×120	**l**	72×60
m	8.1×0.09		

4 Gerald makes toy wooden aeroplanes. He sells the aeroplanes for £3.75 each. On Wednesday, Gerald sells 37 aeroplanes. How much money does he get for the 37 aeroplanes?

5 Calculate, using a mental or written method as appropriate:

a	0.3×0.7	**b**	2.54×6
c	8.57×9	**d**	14.3×2.4
e	41.5×3.7	**f**	18.7×5.6
g	29.8×7.4	**h**	69.2×0.67
i	12.7×0.43 m		
j	53.8 km $\times 0.57$ km		
k	39.2 kg $\times 0.98$		
l	34% of 29.6 tonnes		
m	13% of 17.8m		

6 a What is the cost of 64.2 m of wood at £0.87 per metre?

b What is the cost of 31.6 kg of gravel at £0.67 per kilogram?

7 a Kitchen worktops costs £212 per square metre. How much would it cost to cover this area of worktop?

0.4 m

2.3 m

b Horace decides to build a rectangular path in his garden. His path will be 0.82 m wide and 12.6 m long. What is the area of the path?

8 Puzzle

Answer each of the multiplications and find the appropriate letter.
The letters will reveal the answer to the question 'How do you stop moles digging holes in your garden?'

a	4.6×3.8	48.6	= I
b	2.7×18	115.94	= T
c	23.6×0.78	0.4704	= V
d	387×4.6	38.896	= H
e	187×0.62	106.68	= O
f	34.7×0.8	0.09728	= L
g	73×2.65	18.408	= D
h	29.4×3.8	27.76	= H
i	88.4×0.44	17.48	= H
j	381×0.28	111.72	= S
k	0.56×0.84	2.4548	= E
l	3.23×0.76	1780.2	= E
m	0.256×0.38	193.45	= E

9 Challenge

a Calculate these using a written method:
  i 6.2×2.35
  ii 3.75×2.62
  iii 13.9×7.62
  iv 12.3×6.942
  v $3.42 \times 45.2 \times 17.8$

b Make up a decimal multiplication of your own. Check it using a calculator.

This spread will show you how to:
- ▶▶ Divide by decimals by transforming to division by an integer.
- ▶▶ Make and justify estimates and approximations of calculations.
- ▶▶ Check results using appropriate methods.
- ▶▶ Use known facts to derive unknown facts.

KEYWORDS

Remainder	Approximate
Divisor	Rounding
Dividend	

You can use equivalent calculations to divide decimals.
You make the denominator into an integer by multiplying.

> Always find an approximate answer first.

example

Calculate $0.723 \div 5.8$ (to 3 dp).

Approximate: $0.723 \div 5.8 \approx 0.72 \div 6 = 0.12$

Write the division as a fraction:

$$0.723 \div 5.8 = \frac{0.723}{5.8} = \frac{7.23}{58}$$

$\times 10$ (top), $\times 10$ (bottom)

> Multiply the dividend and the divisor by the same power of 10 to make the divisor an integer.

Now divide:

```
58) 7.2300
   -5.8000        58 × 0.1 = 5.8
    1.4300
   -1.1600        58 × 0.02 = 1.16
    0.2700
   -0.2320        58 × 0.004 = 0.232
    0.0380
   -0.0348        58 × 0.0006 = 0.0348
    0.0032
```

> $7.23 \div 58$ and $0.723 \div 5.8$ are the same calculation.

$7.23 \div 58 = 0.1246$ remainder 0.0032
$0.723 \div 5.8 = 0.125$ (3 decimal places)

It is important to think about the effect of rounding on the accuracy of the final answer when dividing:

$6.61 \div 2.45 = 2.698$ (to 3 dp)

Rounding to the nearest whole number gives: $7 \div 2 = 3.5$
Rounding each number down gives: $6 \div 2 = 3$

> In this case rounding both numbers up or down is better because the ratio of the numbers is more accurately maintained.

Exercise N2.7

1 Calculate, giving your answer to 1 dp where appropriate:
   a 0.6 × 11   b 0.09 × 15
   c 0.07 × 20   d 3.2 × 9
   e 71.4 ÷ 0.1   f 899 ÷ 29
   g 589 ÷ 34   h 167.5 ÷ 19
   i 115 ÷ 2.5   j 288 ÷ 6.4
   k 141.4 ÷ 43   l 51 ÷ 2.6

2 a At Cheapo Supermarket, tins are stacked in layers inside cupboards. The height of a tin is 14 cm. The height of a cupboard is 1.34 m. How many layers of tins can be kept in the cupboard?
   b Glyn has to put sheets of manuscript into folders. Each folder can hold exactly 24 pieces of manuscript. Glyn has 500 sheets of manuscript. How many folders can be completely filled with sheets of manuscript?
   c At Gro-well Garden Centre, shrubs cost £23 each. Mr Titchmarsh has £300 to spend on shrubs. How many shrubs can he buy?

3 Calculate these using a written method, giving your answer to 2 dp where appropriate:
   a 129 ÷ 0.6   b 406.8 ÷ 0.9
   c 236.4 ÷ 0.08   d 0.056 ÷ 0.0037
   e 0.724 ÷ 8.4   f 4.29 ÷ 0.056
   g 12.8 ÷ 0.65   h 0.387 ÷ 7.2
   i $\frac{3}{5} \div \frac{11}{25}$ j $\frac{14}{15} \div \frac{9}{7}$

4 a Sheila has to shovel sand into bags. Each bag can hold 0.8 kg of sand. If Sheila has 140 kg sand, how many bags will she need?
   b A piece of fabric is 135.3 m long. A machine cuts the fabric into strips of length 0.06 m. How many strips can be cut out of a 135.3 m piece of fabric?

5 a Angus is driving along a motorway at a speed of 1.2 km per minute. He has to travel another 324.6 km at this speed before he leaves the motorway. How long will it take Angus to travel 324.6 km at 1.2 km per minute?
   b Harry buys his biodegradable washing-up liquid in bulk. He is not sure how much his container can hold, but he can fill it for a cost of £9.72. The washing-up liquid costs £1.35 a litre. How much liquid can Harry's container hold?

6 **Investigation**
   Here are the lengths of some creatures:

Creature	Length
Blue whale	26.2 m
Polar bear	2.45 m
Human male	1.79 m
Rusty spotted cat	39 cm
Pygmy jerboa	0.036 m
Bumblebee bat	0.029 m
Scorpion	0.013 m
Dwarf pygmy goby fish	0.0086 m
Micro-moth	2 mm

   a How many scorpions would fit along the length of a polar bear?
   b How many times bigger than a pygmy jerboa is a human male?
   c What percentage of the length of a blue whale is a polar bear?
   d Which creature is just over 200 times the length of another creature?
   e Make up some more statements of your own using the table of information.

This spread will show you how to:
▶▶ Know that a recurring decimal is an exact fraction.
▶▶ Use a calculator efficiently and appropriately.
▶▶ Use a range of function keys on the calculator.
▶▶ Enter numbers and interpret the display.

KEYWORDS
Brackets   Product
Adjacent
Recurring decimal
Order of operations

You can use a calculator to help solve complex calculations.

example

Calculate 3×4.3^2

Input: [3] [×] [4] [.] [3] [x^2] [=]

Answer = 55.47

Scientific calculators use the order of operations.
The calculator works out 4.3^2 and then multiplies the answer by 3.

Division calculations may contain hidden brackets.

example

Calculate $\dfrac{12.49 \times 3.67}{2.3 \times (4.65 - 2.7)}$

Using brackets: $\dfrac{12.49 \times 3.67}{(2.3 \times (4.65 - 2.7))}$

Input: [1] [2] [.] [4] [9] [×] [3] [.] [6] [7] [÷]

[(] [2] [.] [3] [×] [(] [4] [.] [6] [5] [−] [2] [.] [7] [)] [)] [=]

Answer = 10.220356...
= 10.2 (1 dp)

The calculator divides the product of 12.49 and 3.67 by the result of 2.3 × (4.65 − 2.7).

You may need to input a multiplication sign for adjacent brackets.

example

Calculate $(12.4 - 6.37)(2.9 - 4.23)$

Use a multiplication sign: $(12.4 - 6.37) \times (2.9 - 4.23)$

Input: [(] [1] [2] [.] [4] [−] [6] [.] [3] [7] [)]

[×] [(] [2] [.] [9] [−] [4] [.] [2] [3] [)]

Answer = ⁻8.0199
= ⁻8.0 (1 dp)

The calculator multiplies the contents of the brackets together.

Some fractions result in recurring decimals.

Input: 1 [a^b/c] 9

Press [a^b/c] again to convert to a decimal: $\frac{1}{9} = 0.111111...$

Press [a^b/c] again to convert back.

$\frac{2}{9} = 0.2222...$
$\frac{3}{9} = 0.3333...(= \frac{1}{3})$
$\frac{4}{9} = 0.4444...$
$\frac{5}{9} = 0.5555...$
$\frac{6}{9} = 0.6666...(= \frac{2}{3})$

Exercise N2.8

1 a Write down an approximate answer for each of these questions, clearly showing your method of approximation.

i 29^2	**ii** $\sqrt{70}$
iii 6.8^2	**iv** $\sqrt{523.9}$
v $\frac{4}{9} + \frac{3}{7} + \frac{17}{21}$	**vi** $2\frac{5}{6} + 3\frac{4}{7} - \frac{9}{14}$

b Use your calculator to work out the exact answers, where appropriate giving your answer to two decimal places or leaving your answer as a fraction.

2 Solve these calculations using a mental, written or calculator method. Where appropriate, give your answer to 2 dp.

a $(19.1 - 1.9)(19.1 + 1.9)$

b $\dfrac{(3.7 - 6)^2 + 2.8}{\sqrt{(2.5^2 + 7 \times 3.4)}}$

c $\dfrac{44.87 \times 72.1}{2.04 \times (7.1 - 3.8)}$

d $\dfrac{7 \times \sqrt{(7.3^2 + 8^2)}}{3.5^2 - 1.7}$

e $[(4.3)^2 - 3(8.1 - 0.076)]^2$

f $7.3 + (4.8 - \sqrt{(11^2 - 6.4^2)})$.

3 a Find 5% of 15% of 2440 m.

b Calculate $\frac{3}{8}$ of $\frac{2}{5}$ of £23.

c An antique vase increases in value by 13% per year. Its original value is £150. Calculate the value of the vase after 5 years.

4 Puzzle

Four consecutive whole numbers have a product of 863 040.

What are the numbers?

5 Investigation

Charlie makes concrete blocks. To work out the volume of water (in litres) he needs to make a concrete block, he uses this rule:

> Volume of water = (length of block in metres)^2 + 6 (length of block in metres)

a Calculate the volume of water in litres needed to make a concrete block of length 4.86 m.

b Calculate the length of block that can be made using 100 litres of water. Give your answer to two decimal places.

6 Use your calculator to evaluate each of these formulae, giving your answers to 1 dp as appropriate:

a Find the area (A) of a trapezium if $A = \frac{1}{2}(a + b)h$ when $a = 2.4$ m, $b = 3.87$ m and $h = 2.7$ m.

b Find the circumference (C) of a circle if $C = \pi d$, when $d = 34.7$ cm and $\pi = 3.14$ (2 dp).

c Find the area (A) of an annulus if $A = \pi(r_1^2 - r_2^2)$, when $r_1 = 18\frac{1}{2}$ cm and $r_2 = 11\frac{2}{5}$ cm. (Use $\pi = \frac{22}{7}$)

d Find the area (A) of a triangle if $A = \sqrt{(s(s - a)(s - b)(s - c))}$, when $a = 6.4$ cm, $b = 4.2$ cm, $c = 8.9$ cm and $s = \frac{1}{2}(a + b + c)$.

e Find the value of b if $b = \frac{3p^2(p - 5)}{4p}$ when $p = 3.8$.

7 Puzzle

a By substituting values of your own choice, check to see which of these formulae are the same (use $\pi = 3.14$):

$V = \pi r^2 h$	$h = \frac{\pi}{r^2 V}$
$r = \sqrt{(\frac{V}{\pi h})}$	$V = \frac{\pi}{r^2 h}$

b Explain and justify your answers.

You should know how to ...

1 Make and justify estimates and approximations of calculations.

2 Use standard column procedures to add and subtract integers and decimals of any size.

3 Multiply and divide by decimals, dividing by transforming to division by an integer.

4 Give solutions to an appropriate degree of accuracy.

Check out

1 a I pay £24 a week to travel to work on the bus. I work for 50 weeks in a year. How much does it cost to travel to work on the bus in a year?

b I can buy a special bus pass that lasts for 12 weeks and costs £252. How much is that per week?

2 Estimate the answer to each of these calculations, giving your answer to an appropriate degree of accuracy.

a $4.08 \times (3.86 \div 1.29)$

b $\dfrac{97 \times 0.78}{\sqrt{26.1}}$

3 Find the perimeter of this shape.

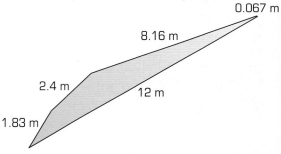

0.067 m
8.16 m
2.4 m
12 m
1.83 m

4 Without using a calculator:

a Calculate the value of 22% of £58.40.

b Calculate how many bags of rice (each weighing 0.7 kg) there are in a box of rice weighing 161 kg.

Graphs of functions

This unit will show you how to:

▶▶ Extend knowledge of integer powers of 10.

▶▶ Use the prime factor decomposition of a number.

▶▶ Use the function keys for powers.

▶▶ Use index notation for integer powers and simple instances of the index laws.

▶▶ Simplify or transform expressions by taking out single factors.

▶▶ Given values for m and c, find the gradient of the lines given by equations of the form $y = mx + c$.

▶▶ Construct functions arising from real-life problems and plot their corresponding graphs.

▶▶ Interpret graphs arising from real situations, including distance–time graphs.

▶▶ Solve increasingly demanding problems and evaluate solutions.

▶▶ Solve substantial problems by breaking them into simpler tasks.

▶▶ Generate points and plot graphs of linear functions.

▶▶ Represent problems and interpret solutions in algebraic and graphical forms.

▶▶ Move from one form to another to gain a better perspective on the problem.

The bigger the gradient, the steeper the hill.

Before you start

You should know how to ...

1 Use index notation.
$t \times t = t^2$

2 Find the prime factors of a number.
$12 = 2 \times 6 = 2 \times 2 \times 3$

3 Substitute values into formulae.

Check in

1 Write using index notation:

a $3 \times 3 \times 3$ b $m \times m \times m \times m$

c $f \times f \times f \times f \times f \times f \times f$

2 Find the prime factors of:

a 20   b 24   c 32   d 100

3 Copy and complete the table for given values of x when $y = 4x - 3$.

x	−2	−1	0	1	2
$y = 4x - 3$		−7			

Finding factors

This spread will show you how to:
- ▶▶ Recognise and use multiples, factors and primes.
- ▶▶ Use squares and cubes and corresponding roots.
- ▶▶ Use index notation for small positive integer powers.

KEYWORDS

Prime number
Square number
Factor
Prime factor decomposition

Every number can be made up of factor pairs.

$17 = 1 \times 17$ The factors of 17 are 1 and 17.

$22 = 1 \times 22$
$22 = 2 \times 11$ The factors of 22 are 1, 2, 11 and 22.

$36 = 1 \times 36$
$36 = 2 \times 18$
$36 = 3 \times 12$
$36 = 4 \times 9$
$36 = 6 \times 6$ The factors of 36 are 1, 2, 3, 4, 6, 9, 12, 18 and 36.

> ▶ A number with only two factors is a **prime** number.
> 17 is a prime number.

> ▶ A number with an odd number of factors is a **square** number.
> 36 is a square number.
> The factors in one of the pairs are exactly the same: **6 × 6**.

You can write a number as the product of prime factors.
The tree method is useful.

example

a Complete a factor tree for 84.
b Write 84 as a product of its prime factors.
....
a Break down the factors until they are all prime numbers:

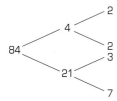

b You write $84 = 2 \times 2 \times 3 \times 7$
$\qquad\qquad\quad = 2^2 \times 3 \times 7$

$2^2 \times 3 \times 7$ is the prime factor decomposition of 84.

Exercise A4.1

1 In a productagon, the number in the square is the product of the numbers in the circles on either side.
Copy and complete these productagons.

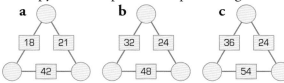

2 Find all the factor pairs of each of these numbers.
  a 18   **b** 15   **c** 28
  d 80   **e** 36

3 **a** Find all the factors of 42 and of 70. Write the factors in separate lists.
  b List the common factors.
  c Write down the highest common factor of 42 and 70.

4 Complete prime factor trees for each of these numbers. The first one is started for you.

a
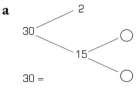
30 =

  b 16   **c** 40   **d** 56
  e 100   **f** 38   **g** 108

5 7^3 means $7 \times 7 \times 7 = 343$
Write down the actual value of each of these:
  a 6^2 **b** 8^2 **c** 12^2
  d 30^2 **e** 3^3 **f** 1^3
  g 5^3 **h** 9^3

6 $\sqrt{60}$ lies somewhere between 7 ($\sqrt{49}$) and 8 ($\sqrt{64}$).
Write down the whole numbers that each of these square roots lie between.
  a $\sqrt{30}$ **b** $\sqrt{10}$
  c $\sqrt{75}$ **d** $\sqrt{42}$
  e $\sqrt{110}$

7 **Mix and match**
The prime factors in group A match a number in group B.
Match each pair together.

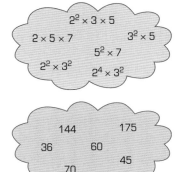

8 The number 48 has prime factors $2^4 \times 3$ (which is $2^2 \times 3 \times 2^2$).
  a Find four more numbers that include $2^2 \times 3$ as prime factors.
  b What is the largest number less than 100 that includes the prime factors $2^2 \times 3$?

9 **Investigation**
You can write 13 as the difference of two consecutive squares:
$7^2 - 6^2 = 49 - 36 = 13$.
What other numbers less than 30 can be written as a difference of two consecutive squares?

Using prime factors

This spread will show you how to:
- ▶▶ Use the prime factor decomposition of a number.
- ▶▶ Simplify or transform algebraic expressions by taking out common factors.

KEYWORDS
Highest common factor
Lowest common multiple
Prime factor

For any two or more numbers:

> ▶ The highest common factor (HCF) is the largest number that each of the numbers divide by exactly.

> ▶ The lowest common multiple (LCM) is the smallest number that each of the numbers divide into exactly.

You can find the HCF and LCM using prime factors.

example

Find the HCF and LCM of the two numbers 24 and 30.

Factors of 24

Factors of 30

For the HCF, find all the factors common to both numbers:

$2 \times 2 \times ②\times ③$ and $②\times③\times 5$

HCF $=$ 2×3 $= 6$

For the LCM, use the factors of the first number and any new factors in the second number.

$(2 \times 2 \times 2)\times③$ and $2 \times 3 \times ⑤$

LCM $= 2 \times 2 \times 2 \times 3 \times 5 = 120$

You can use the LCM to help order fractions.

example

Order these fractions from largest to smallest: $\frac{5}{8}$ $\frac{7}{12}$ $\frac{13}{20}$

Find the LCM of 8, 12 and 20: $8 = (2 \times 2 \times 2)$ $12 = 2 \times 2 \times ③$ $20 = 2 \times 2 \times ⑤$

LCM $= 2 \times 2 \times 2 \times 3 \times 5 = 120$

Use 120 as the common denominator:
$\frac{5}{8} = \frac{5 \times 15}{8 \times 15} = \frac{75}{120}$
$\frac{7}{12} = \frac{7 \times 10}{12 \times 10} = \frac{70}{120}$
$\frac{13}{20} = \frac{13 \times 6}{20 \times 6} = \frac{78}{120}$

In order they are $\frac{78}{20}$ then $\frac{75}{120}$ then $\frac{70}{120}$, so $\frac{13}{20}$ then $\frac{5}{8}$ then $\frac{7}{12}$.

Exercise A4.2

1 a Copy and complete these factor trees for 70 and 112:

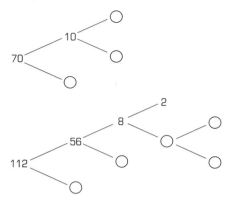

b Write each number as a product of prime factors.

c Use the prime factors to work out the HCF and LCM.

2 a Find the prime factors of these pairs of numbers:
  i 45 and 81   **ii** 54 and 126
  iii 80 and 104

b Find the HCF.

c Find the LCM.

3 Find the LCM of the denominators of these fractions and write down which of the fractions in each pair is larger.

a $\frac{5}{9}$ and $\frac{8}{15}$ **b** $\frac{7}{12}$ and $\frac{12}{21}$

c $\frac{3}{8}$ and $\frac{5}{14}$ **d** $\frac{11}{28}$ and $\frac{16}{35}$

4 Find the HCF of each term in these expressions.
Use the HCF to factorise the expressions.
The first one is started for you.

a $12x + 15y = 2 \times 2 \times 3 \times x + 3 \times 5 \times y$
$$= 3(2 \times 2 \times x \quad + \qquad)$$
$$= 3(\qquad + \qquad)$$

b $20r - 16p$ **c** $28a + 21b$

d $18d - 45e$ **e** $36f + 24g$

f $49x + 21y - 14z$

5 The HCF of each pair of terms is written in the cloud.

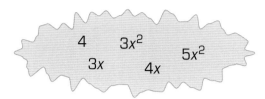

Match each pair with its HCF.

a $3x^2$ and $6x$

b $12x$ and $4x^3$

c $5x^3$ and $10x^2$

d $9x^3$ and $6x^2$

e $20x$ and 8

6 Find the HCF for each pair of terms.
The first one is done for you.

a $3x^2y$ and $6xy^2$
$3x^2y = 3 \times x \times x \times y$
$6xy^2 = 2 \times 3 \times x \times y \times y$
$HCF = 3 \times x \times y = 3xy$

b $12xy^3$ and $4x^3y$

c $5x^3y^2$ and $10x^2y^3$

d $9x^3y^3$ and $6x^2y$

e $20xy^4$ and $8x^3y^2$

7 Factorise these expressions.
The first one is started for you.

a $3x^2y - 12xy = 3xy(x - \quad)$

b $4x - 6xy$

c $4y^2 + 8y$

d $15xy - 10y^2$

e $14x^2y + 10xy$

f $6xy^2 + 9x^2y$

A4.3 Index notation

This spread will show you how to:
- ▶▶ Extend knowledge of integer powers of 10.
- ▶▶ Use the function keys for powers.
- ▶▶ Use index notation for integer powers and simple instances of the index laws.

KEYWORDS

Index	Indices
Base	Cubed
Power	Index laws
Squared	

6^2 means 6×6.

You say '6 squared' or '6 to the power of 2'.

5^3 means $5 \times 5 \times 5 = 125$.

You say '5 cubed' or '5 to the power of 3'.

In $\quad 5^3 \longrightarrow$ 3 is the index or power.
$\quad\quad\quad\quad\quad\longrightarrow$ 5 is the base.

You use this button on a scientific calculator to work out the value of an index: $\boxed{x^y}$

To work out 5^3 you press $\boxed{5}\ \boxed{x^y}\ \boxed{3}\ \boxed{=}$ answer is 125

To work out 3^5 you press $\boxed{3}\ \boxed{x^y}\ \boxed{5}\ \boxed{=}$ answer is 243

You can multiply and divide expressions with the same bases.

$$6^2 \times 6^3 = 6 \times 6 \quad \times \quad 6 \times 6 \times 6 \quad = 6^5$$
$$= 36 \quad\quad \times \quad\quad 216 \quad\quad = 7776$$

$$3^5 \div 3^2 = (3 \times 3 \times 3 \times 3 \times 3) \div (3 \times 3) = 3 \times 3 \times 3 \quad = 3^3 \quad\quad = 27$$

You use the same letter to show the same base in algebra.

$$n^5 \times n^3 = n \times n \times n \times n \times n \quad \times \quad n \times n \times n \quad\quad = n^8$$

$$n^5 \div n^3 = (n \times n \times n \times n \times n) \quad \div \quad (n \times n \times n) \quad\quad = n \times n \quad = n^2$$

You can generalise these index laws for any powers.
Use different letters to stand for different powers.

To multiply powers of the same base, you add the indices.

> ▶ $x^a \times x^b = x^{(a+b)}$
> For example, $x^5 \times x^7 = x^{(5+7)} = x^{12}$

To divide powers of the same base, you subtract the indices.

> ▶ $x^a \div x^b = x^{(a-b)}$
> For example, $x^7 \div x^5 = x^{(7-5)} = x^2$

Exercise A4.3

1 Use the $\boxed{x^y}$ button on your calculator to work out each of these:

   a 6^4 **b** 3^7
   c 4^5 **d** 2^9
   e 5^5 **f** 9^3
   g 8^6 **h** 15^3
   i 7^5

2 Use your calculator to work out the value of x in these equations:

   a $4096 = 4^x$ **b** $729 = 3^x$
   c $6561 = 9^x$ **d** $117\,649 = x^6$
   e $1024 = x^{10}$ **f** $65\,536 = x^8$
   g $x = 6^7$ **h** $390\,625 = x^8$

3 Copy and complete these tables. Use your calculator if necessary.

a

Index form	Value
10^4	
10^3	1000
10^2	
10^1	
10^0	
10^{-1}	
10^{-2}	

b

Index form	Value
2^4	
2^3	8
2^2	
2^1	
2^0	
2^{-1}	
2^{-2}	

Hint: to input ⁻3 press $\boxed{3}$ $\boxed{+/-}$

4 Write each of these expressions as a power with one base.
The first one is done for you.

   a $3^2 \times 3^4 = 3 \times 3 \times 3 \times 3 \times 3 \times 3 = 3^6$
   b $4^3 \times 4^2$ **c** $5^6 \times 5^2$
   d $6^5 \times 6^3$ **e** $7^4 \div 7^2$
   f $9^3 \div 9^2$ **g** $10^5 \div 10^2$
   h $2^7 \div 2^3$

5 Write each of these expressions as a single power in the form y^n:

   a $y^3 \times y^2$ **b** $y^5 \times y^2$ **c** $y^7 \times y^3$
   d $y^9 \div y^4$ **e** $y^3 \div y^2$ **f** $y^3 \div y$
   g $y^6 \div y^6$ **h** $y^2 \div y^3$ **i** $y^3 \div y^5$

6 Challenge
Copy the table. Decide which statements are always true, sometimes true or never true and put them in the correct columns in your table.

Always	Sometimes	Never

   a $7^x = 1$ **b** $x^5 \div x^3 = x^2$
   c $x^2 + x^2 = x^4$ **d** $x^2 \times x^3 = x^6$
   e $x^0 = 10$ **f** $x^2 < x$
   g $2^x = x^2 \ (x \neq 2)$

For each statement that is sometimes true, give a value for x to show that it is true.

7 Challenge
   a Write $y \times y \times y \times y$ in its simplest form.
   b What is $x^y + y^x$ when $x = 2$ and $y = 3$?
   c If $2^x + 3 = 4$, what is x?
   d If $x^x = 27$, what is x?

Solving problems involving indices

This spread will show you how to:
▶▶ Use the function keys for powers.
▶▶ Use index notation for integer powers and simple instances of the
   index laws.

KEYWORDS

Power   Sequence
Integer   Digit

The powers of 7 are: 7^1 7^2 7^3 7^4 7^5 7^6 ...
The values are: 7   49   343   2401   16 807   117 649   ...

Each number is odd.
You can explain why: 7 is odd.
  $7^2 = 7 \times 7 = \text{odd} \times \text{odd} = \text{odd}$
  $7^3 = 7^2 \times 7 = \text{odd} \times \text{odd} = \text{odd}$
  etc.

Notice that $7^4 = (7 \times 7) \times (7 \times 7)$ $= 7^2 \times 7^2 = 49^2$
  $7^6 = (7 \times 7 \times 7) \times (7 \times 7 \times 7)$ $= 7^3 \times 7^3 = 343^2$
  $7^8 = (7 \times 7 \times 7 \times 7) \times (7 \times 7 \times 7 \times 7)$ $= 7^4 \times 7^4 = 2401^2$

The last digits of each of the values form a sequence:

7   9   3   1   7   9 ...

You can show why the sequence repeats by looking at multiples of 7:

  $7 \times 7 = 49$ $9 \times 7 = 63$ $3 \times 7 = 21$ $1 \times 7 = 7$ $7 \times 7 = 49$ $9 \times 7 = ...$

You can prove some results in algebra by first looking at examples.

example

Is the result of $2^n + 3^m$ always even, always odd or can't you tell?
(n and m are integers $\geqslant 1$)

Try some simple cases:

$2^1 = 2$ $3^1 = 3$
$2^2 = 2 \times 2 = 4$ $3^2 = 3 \times 3 = 9$
$2^3 = 2 \times 2 \times 2 = 8$ $3^3 = 3 \times 3 \times 3 = 27$
$2^4 = 2 \times 2 \times 2 \times 2 = 16$ $3^4 = 3 \times 3 \times 3 \times 3 = 81$
$2^5 = 2 \times 2 \times 2 \times 2 \times 2 = 32$ $3^5 = 3 \times 3 \times 3 \times 3 \times 3 = 243$

2^n is always even × even = even 3^n is always odd × odd = odd

$2^n + 3^m$ is always even + odd = odd

$2^n + 3^m$ is always odd.

Exercise A4.4

1 Choose one of the signs $>$, $<$ or $=$ to make each of these statements true.
  a $8^5 \_\_ 8^4$ **b** $4^3 \_\_ 3^4$
  c $2^4 \_\_ 4^2$ **d** $5^3 \_\_ 3^5$
  e $4^3 \_\_ 2^6$ **f** $8^3 \_\_ 2^7$
  g $2^3 \times 5^3 \_\_ 10^3$ **h** $(0.5)^2 \_\_ (0.5)^3$

2 Work out the value of n in each of these equations.
  a $3^3 \div 3^2 = 3^n$ **b** $5^4 \div 5^2 = 5^n$
  c $9^6 \div 9^2 = 9^n$ **d** $8^3 \times 8^2 = 8^n$
  e $5^4 \times 5^3 \div 5^2 = 5^n$ **f** $4^6 \div 4^n = 4^3$

3 a Calculate the values of the terms of this sequence:

     $4^1 \quad 4^2 \quad 4^3 \quad 4^4 \quad 4^5 \quad 4^6$

  b What digit does 4^{12} end in?
  c What digit does 4^{13} end in?
  d Will 4^n ever be an odd number? Explain your answer.

4 You are told that the last digit for the value of 5^x is 7.
Could this be a true statement? Explain your answer.

5 a Match each number in the box to its equivalent index form without using a calculator.

2187	3^a	1 000 000
	4^b	
	5^c	7776
78 125	6^d	262 144
	10^e	

  b Use a calculator to work out the values for a, b, c, d and e.

6 Look at these numbers:

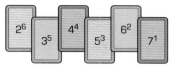

  a Which is the largest?
  b Which is equal to 4^3?
  c Which two add up to 10^2?

7 Which of these are square numbers?

$2^3 \quad 2^4 \quad 2^5 \quad 2^6 \quad 2^7$

8 Copy and complete these statements:
  a If $2^8 = 256$ then $2^9 = \_\_\_$
  b If $2^{15} = 32\,768$ then $2^{14} = \_\_\_$
  c If $64 = 8^n = 4^p = 2^q$ then
     $n = \_\_\_$, $p = \_\_\_$, $q = \_\_\_$
  d If $\sqrt{144} = (144)^{\frac{1}{2}} = 12$
     then $(81)^{\frac{1}{2}} = \_\_\_$

9 Triples
The numbers in the cloud are in triples.

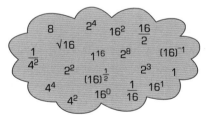

For example: $\frac{16}{2} = 8 = 2^3$.
Find five more triples.

10 You are told that $x^2 - y^2 = (x + y) \times (x - y)$
  a Check the formula is true when $x = 7$ and $y = 5$ (so $7^2 - 5^2$).
  b Use the formula to work out the values of $17^2 - 13^2$ and $56^2 - 44^2$.
     Check your answers using a calculator.

This spread will show you how to:

▶▶ Generate points in all four quadrants and plot the graphs of linear functions, where y is given explicitly in terms of x.

▶▶ Recognise that equations of the form $y = mx + c$ correspond to straight-line graphs.

KEYWORDS

Linear function
Straight-line graph
Gradient　　　Intercept
Coordinate pair

You draw the graph of a function by substituting values to generate coordinate pairs.

To draw the graph of: $y = 2x + 3$,

▶ choose x values　　　($^-2$,　) ($^-1$,　) (0,　) (1,　) (2,　)
▶ work out y values.　　($^-2$, $^-1$) ($^-1$, 1) (0, 3) (1, 5) (2, 7)

Draw a suitable grid.

▶ The x-axis will go from $^-2$ to 2.
▶ The y-axis will go from $^-1$ to 7.

You could make a table of values:

x	$^-2$	$^-1$	0	1	2
$y = 2x + 3$	$^-1$	1	3	5	7

Plot the points and join them.

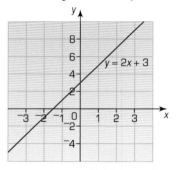

The table of values shows that when the x values go up in 1s the y values go up in 2s.
The **gradient** is 2.

The graph cuts the y-axis at $y = 3$.
The **y-intercept** is 3.

▶ $y = 2x + 3$ is a linear function. It produces a straight-line graph.
▶ The gradient is 2 and the y-intercept is 3.

example

a Draw the graph of $y = 4x - 3$ for x values from $^-3$ to 3.
b Find the y-intercept and gradient of the graph.

a Make a table of values then plot the graph:

x	$^-2$	$^-1$	0	1	2
$y = 4x - 3$	$^-11$	$^-7$	$^-3$	1	5

b The graph cuts the y-axis at $^-3$.
The y-intercept is $^-3$.
When the x value goes up by 1, the y value goes up by 4.
The gradient is 4.

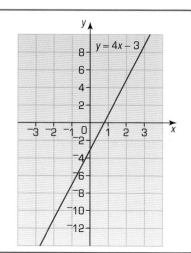

Exercise A4.5

1 Copy the coordinate grid.

  a Plot the points:

    (3, 9) (⁻1, ⁻5), (⁻5, ⁻7) (4, 10)
    (⁻2, ⁻8) (⁻2, ⁻1) (2, 4) (1, 5)

The points lie on two different lines.

  b Draw the two straight lines.

  c For each line, write the coordinate pair when $x = 0$, $(0, ?)$.

  d For each line, write the coordinate pair when $y = 7$, $(?, 7)$.

  e Label the steeper line A.

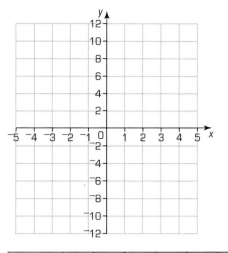

2 **a** Copy and complete these tables of values for the linear functions $y = 2x + 1$ and $y = 3x - 2$.

  b Plot the points on a coordinate grid.

  c Write down the coordinates of the point where the two lines intersect.

x	-3	-2	-1	0	1	2	3
$y = 2x + 1$			-1				

x	-3	-2	-1	0	1	2	3
$y = 3x - 2$			-5				

3 **Match**

  a Match the description of the graph to its equation. Use the graph to help you.

The y-coordinate:
- is always double the x-coordinate
- is always equal to the x-coordinate
- is always a quarter of the x-coordinate
- is always four times the x-coordinate
- is always half of the x-coordinate

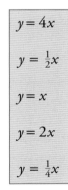

$y = 4x$

$y = \frac{1}{2}x$

$y = x$

$y = 2x$

$y = \frac{1}{4}x$

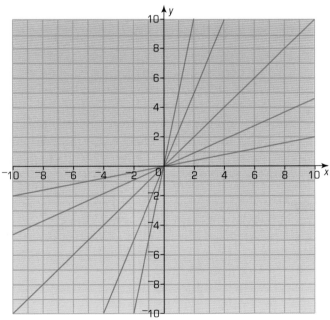

  b Each line goes through (0, 0). Explain how you know this from the equation of each graph.

  c Which line is the steepest (largest gradient)? Which has the smallest gradient?

  d Explain how you find the gradient from the equation.

  e Does the line $y = 7x$ have a large or small gradient?

Finding the gradient

This spread will show you how to:
▶▶ Generate and plot graphs of linear functions.
▶▶ Given values for m and c, find the gradient of lines given by equations of the form $y = mx + c$.

KEYWORDS
Gradient
Intercept
Increase

▶ For a linear function:
   ▶ The gradient of the graph shows the steepness of the line.
   ▶ The y-intercept of the graph shows where it crosses the y-axis.

You can find the gradient and y-intercept from a graph.

Line A crosses the y-axis at $(0, 4)$.
The y-intercept is 4.

To work out the gradient:
When you go across 1
... you need to go up 2.

The gradient is 2.

The equation is $y = 2x + 4$.

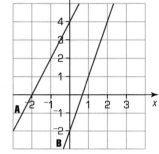

Line B crosses the y-axis at $(0, {}^-2)$.
The y-intercept is ${}^-2$.

To work out the gradient:
When you go across 1
... you need to go up 3.

The gradient is 3.

The equation is $y = 3x - 2$.

▶ The general equation of a line is $y = mx + c$
   where m is the gradient and c is the y-intercept.

In $y = 5x + 1$, the gradient is 5 and the y-intercept is 1.

You can find the gradient from any two points.
You want to know the change in y for every increase in x.

This is a simple ratio: $\text{Gradient} = \dfrac{\text{change in } y}{\text{change in } x}$

example

Find the gradients of these two lines.

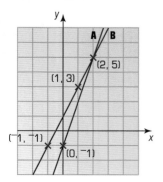

For line A:
the increase in x is from 0 to 2:
$2 - 0 = 2$
the increase in y is from ${}^-1$ to 5:
$5 - {}^-1 = 6$
The gradient $= \frac{6}{2} = 3$.

For line B:
the increase in x is from ${}^-1$ to 1:
$1 - {}^-1 = 2$
the increase in y is from ${}^-1$ to 3:
$3 - {}^-1 = 4$
The gradient $= \frac{4}{2} = 2$.

Exercise A4.6

1 Match each of these graphs with the correct gradient, y-intercept and equation.

i	gradient = 2, y-intercept = $^-3$	**A**	$y = 2x + 3$
ii	gradient = 3, y-intercept = 2	**B**	$y = 3x + 2$
iii	gradient = 2, y-intercept = 3	**C**	$y = 2x - 3$
iv	gradient = 3, y-intercept = $^-2$	**D**	$y = 3x - 2$

a 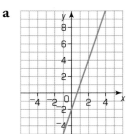   **b** 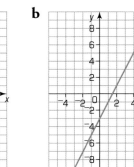   **c** 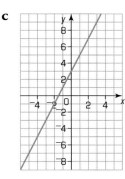   **d**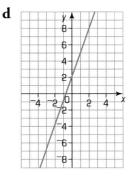

2 For each of these equations:

i	$y = 2x + 6$	**ii**	$y = 3x - 4$
iii	$y = 4x + 1$	**iv**	$y = \frac{1}{2}x - 2$

   a Write down the gradient and y-intercept.
   b Sketch the graph on a coordinate grid with x from $^-3$ to 3 and y from $^-10$ to 10.

3 Work out the gradient of each of these graphs.
Show all your working clearly.

a 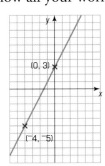   **b** 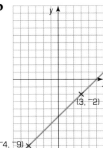   **c** 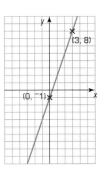   **d**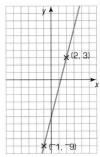

4 Here are the equations of four graphs:

i	$y = x + 3$	**ii**	$y = 3x$
iii	$y = 3x - 2$	**iv**	$y = 2 - 3x$

   a A graph has a gradient of 3 and a y-intercept of $^-2$.
      Which equation matches this description?
   b Write similar descriptions for the other three graphs.

Implicit functions

This spread will show you how to:
▶▶ Generate points and plot graphs of linear functions.
▶▶ Find the gradients of linear functions.

KEYWORDS
Gradient   Inspection
Function (linear)

You can find the gradient of a linear function by inspection (looking).

From an equation: $y = 5x - 2$

$y = ⑤x - 2$ ——→ For every one y you need five x's.
The gradient is 5.

From a graph:

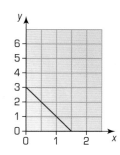

down 3
across $1\frac{1}{2}$

$$\text{Gradient} = \frac{\text{change in } y}{\text{change in } x} = \frac{^-3}{1\frac{1}{2}} = ^-2$$

From a linear function: $2y - 3x + 2 = 0$

$2y - 3x + 2 = 0$
$2y = 3x - 2$
$y = \dfrac{3x - 2}{2} = 1\frac{1}{2}x - 1$ Gradient is $1\frac{1}{2}$ or 1.5

example

Given the linear function $2y - 8x - 6 = 0$

a Rearrange to find the gradient.
b Find the y-intercept of the graph of the function.
c Use this information to sketch the graph.

....

a $2y - 8x - 6 = 0$
  $2y = 8x + 6$
  $y = ④x + 3$ Gradient = 4

b Intercept of $y = 4x + 3$ is 3.

c

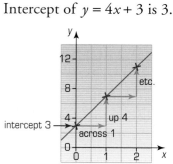

intercept 3
up 4
across 1
etc.

Exercise A4.7

1 Rearrange these implicit functions to make y the subject of the equation:

a $y - 2x = 3$ **b** $y - 3x + 2 = 0$ **c** $y + 3x - 4 = 0$

d $\frac{y}{3} - x = 0$ **e** $\frac{y}{5} - 2x = 0$ **f** $\frac{y}{2} + 6x = 0$

g $2y - 4x = 6$ **h** $2y + 3x = 12$ **i** $2y - 5x - 3 = 0$

2 Match each implicit function with a linear equation and a graph.

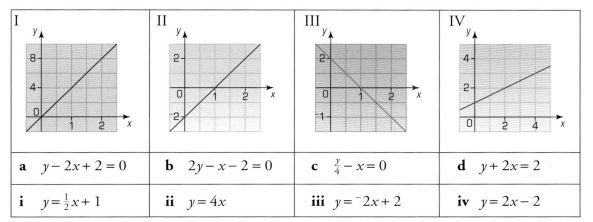

I	II	III	IV
a $y - 2x + 2 = 0$	**b** $2y - x - 2 = 0$	**c** $\frac{y}{4} - x = 0$	**d** $y + 2x = 2$
i $y = \frac{1}{2}x + 1$	**ii** $y = 4x$	**iii** $y = {}^{-}2x + 2$	**iv** $y = 2x - 2$

3 i Match each of these implicit functions with their gradient and intercept.

a $y - 3x + 2 = 0$

b $y - 4x = 5$

c $\frac{y}{4} - x = 0$

d $y + 3x - 2 = 0$

	Gradient	Intercept
i	4	0
ii	3	${}^{-}2$
iii	${}^{-}3$	2
iv	4	5

ii On a grid with x from ${}^{-}4$ to 4, and y from ${}^{-}15$ to 15, sketch each of the functions above.

4 Challenge

Here is a kite ABCD on a grid.

a Match the correct line to these three equations.
- ▸ line through A and B
- ▸ line through C and B
- ▸ line through D and C
- ▸ line through D and A
- ▸ line through D and B
- ▸ line through C and A

$$y = 0$$
$$y + 2x = 4$$
$$y + x + 2 = 0$$

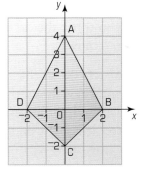

b Work out the equations of the remaining three lines.

Real-life graphs

This spread will show you how to:
▶▶ Construct functions arising from real-life problems and plot their corresponding graphs.
▶▶ Interpret graphs arising from real-life situations.

KEYWORDS
Trend
Horizontal axis
Vertical axis
Interpret

You can show real-life problems on graphs.
The graphical representation is useful for showing trends.

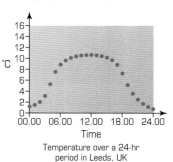

Temperature over a 24-hr period in Leeds, UK

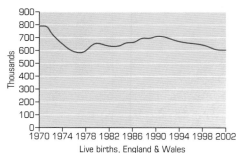

Live births, England & Wales

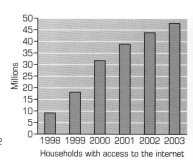

Households with access to the internet

You often need to interpret the graphs carefully.

example

This graph shows the amount of fuel left in a Formula 1 racing car at any time throughout a race.

Use the graph to answer these questions:

a   What do each of the axes show?
b   How much fuel does the car start with?
c   How much fuel did the car have left at the end of the race?
d   At what times during the race does the driver stop to refuel?
e   When do you think the car was travelling at its fastest speed? How do you know?

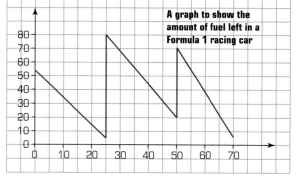

A graph to show the amount of fuel left in a Formula 1 racing car

a   The vertical axis shows the amount of fuel (in litres) left in the car.
     The horizontal axis shows the time taken in the race (in minutes).
b   The car starts with 54 litres of fuel. This is where the graph crosses the fuel axis.
c   It had 6 litres left at the end.
d   The driver stops after 25 minutes and again after 50 minutes to refuel.
     You can tell because the amount of fuel in the tank suddenly increases.
e   A car travels fastest when it has the least amount of fuel in the tank (it is lighter).
     It was probably about 24 minutes into the race, just before it stopped to refuel, and again at the end of the race.

Exercise A4.8

1 The graph shows the height of a sunflower grown from seed.
The main sunflower opened after 12 weeks.

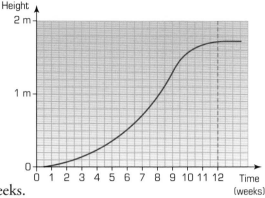

  a Describe how the plant has grown in the 12 weeks.
  b How tall did the sunflower grow?
  c How tall was the sunflower after 7 weeks? How tall was the sunflower after 9 weeks?
  d Estimate how many cm per day the plant grew in the period from 7 weeks to 9 weeks.

2 The table shows the weight of a baby during his first year.

Age	birth	1 month	2	3	4	5	6	7	8	9	10	11	12
Weight/kg	3.2	3.5	4.2	5.1	4.9	6.2	7.3	8.5	9.8	10.5	11	11.4	11.8

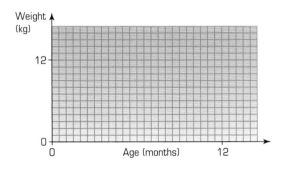

  a Plot this information on a copy of the graph and join the points smoothly.
  b Estimate how old the baby was when he weighed 6 kg.
  c When the baby was 3 months old he had a virus. How is this shown in the graph?
  d A healthy baby should grow to about 3 times their birth weight after a year. Do you think this baby was healthy?

3 This graph shows the relationship between the area of a square and the length of each side.

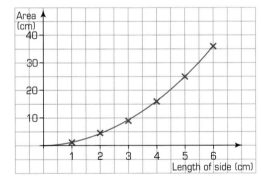

  a What is the area of a square of side length 4 cm?
  b The area of a square is 30 cm^2. Estimate the length of each side.
  c The length of a side of a square is 2.6 cm. Estimate its area to the nearest 1 cm^2.

4 Koi carp grow to be large fish. Here are the lengths of a koi measured during its first 8 years of life.

Age (x)	6 months	1 yr	2 yrs	4 yrs	5 yrs	7 yrs	8 yrs
Length (y)	4 cm	5 cm	8 cm	17 cm	23 cm	38 cm	47 cm

  a Plot these points on a grid and join them up with a smooth line (curve).
  b Estimate from your graph how old the fish would be when it was 30 cm long.
  c Estimate how long the fish was when it was 3 years old.
  d The fish lived to be 10 years old. Estimate how long it was when it died.

133

This spread will show you how to:
- ▶▶ Construct functions arising from real-life problems and plot their corresponding graphs.
- ▶▶ Interpret graphs arising from real-life situations, including distance–time graphs.

KEYWORDS
Distance–time graph
Horizontal axis
Vertical axis

A distance–time graph shows the distance travelled away from a particular place over a period of time.

The steeper the line, the faster the vehicle is travelling.

You can work out the average speed from straight-line graphs:

$$\text{Average speed} = \frac{\text{distance travelled}}{\text{time taken}}$$

Horizontal lines show when the vehicle was stationary (not moving).

The distance travelled is always on the vertical axis

The time taken is always on the horizontal axis

You can tell when a vehicle is accelerating:
A straight line shows a steady, constant speed.

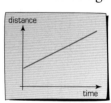

A curved line shows acceleration – the car is travelling at an increasing rate.

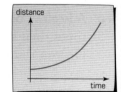

 example

The morning express travels from Swansea to London.
The distance–time graph shows its journey.

a What time does it leave Swansea?
b At what time, and for how long did it first stop?
c How many stations did it stop at?
d When was the train travelling fastest? Explain how you know.

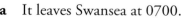

a It leaves Swansea at 0700.
b It first stops at 0745 for 5 minutes, in Cardiff.
c It stops at two stations before London – Cardiff and Bristol.
   It does not stop at Reading as it spends no time in the station.
d The steepest part of the line is between Bristol and Reading, so it was travelling fastest between 0825 and 0855.

Exercise A4.9

1 Karen travels from Manchester to Nottingham by car.
The distance–time graph shows different stages in her journey.
She left Manchester at 16.15 hrs.
Match each of these statements to a stage in the journey.

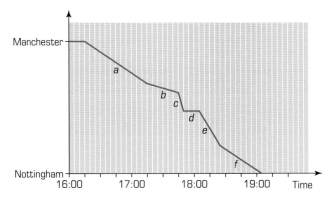

  i She stops at a motorway service station for a break.

  ii She drives quickly on the motorway for 5 minutes.

  iii She drives on the motorway at a constant speed for 1 hour.

  iv She drives quickly on the motorway for 20 minutes.

  v She drives at a slower speed for 40 minutes.

  vi She hits busy traffic on the motorway and drives slowly for 30 minutes.

2 This distance–time graph shows the flight from Earth to the Moon of Apollo 11 in 1969.
It was from this flight that Neil Armstrong and Ed 'Buzz' Aldrin became the first men to step on the moon.

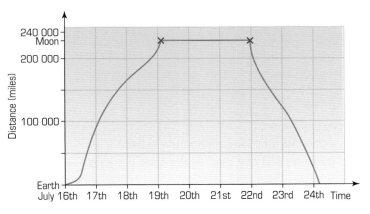

  a Approximately how far is the Moon from the Earth?

  b For how long did Apollo 11 orbit the Moon?

  c How far was it from Earth after 3 days?

  d When were they 100 000 miles from Earth?

  e How far did they travel in day 1? How far in day 2? Can you explain the difference?

3 A funicular railway is often found on the coast, taking passengers up (and down) the side of a cliff.

  a How long does it stop at Midstation?

  b How high is Midstation?

  c What is the speed of the railway in m/s?

  d Describe the journey from sea level to cliff top.

You should know how to …

1 Given values for m and c, find the gradient of the lines given by equations of the form $y = mx + c$.

2 Construct functions arising from real-life problems and plot their corresponding graphs.

3 Interpret graphs arising from real situations.

Check out

1 a Here are two lines:

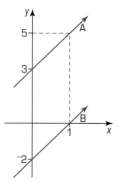

   i What are the gradient and intercept for each line?

   ii Write the equation of each line.

b If the equations of two lines are

   A $y = 3x - 2$ **B** $y = x - 2$

   i What are the gradient and intercept for each line?

   ii At what point do the two lines intersect?

2 This table shows the change in temperature (in °C) of a litre of water as it is boiled and then cools down.

Temperature	20°	100°	85°	75°	67°	50°	36°	26°	20°
Time		1030	1032	1040	1050	1100	1130	1200	1230	1300

Draw a graph with suitable axes to show this information.

3 Use your graph from question 2 to estimate the time when the temperature was:

   a 30 °C

   b 70 °C

   c 55 °C.

   d How long did the water take to cool from 100 °C to 20 °C in total?

Probability

This unit will show you how to:

▶▶ Use the vocabulary of probability in interpreting results involving uncertainty and prediction.

▶▶ Identify all the mutually exclusive outcomes of an experiment.

▶▶ Know that the sum of probabilities of all mutually exclusive outcomes is 1 and use this when solving problems.

▶▶ Estimate probabilities from experimental data.

▶▶ Use efficient methods to add, subtract and multiply fractions.

▶▶ Solve increasingly demanding problems in probability and evaluate solutions.

▶▶ Present a concise, reasoned argument using diagrams and related explanatory text.

There is a 50 : 50 chance of going the right way.

Before you start

You should know how to ...

1 Understand and use the probability scale from 0 to 1.

2 Identify all the possible outcomes of a single event.

3 Convert between fractions, decimals and percentages.

Check in

1 Mark these events on a probability scale:
   a getting a head on a throw of a coin
   b getting a 7 on a throw of a dice.

2 The letters of the word

   C R O C O D I L E

   are placed in a bag.
   List the possible outcomes when you choose a letter at random.

3 Change to percentages:
   a $\frac{3}{4}$ b 0.4   c $\frac{7}{20}$ d 0.35

This spread will show you how to:
▶▶ Use the vocabulary of probability in interpreting results involving uncertainty and prediction.

KEYWORDS

Event	Sample space
Outcome	Favourable
Impossible	Certain
Equally likely	p(n)
Probability scale	

Probability measures the likelihood of an **event** occurring.

It is measured on a scale of 0 to 1.

Impossible   Even chance   Certain

0   0.5   1

Probabilities can be written as fractions, decimals or percentages.

There may be more than one possible **outcome** leading to an event.

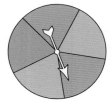

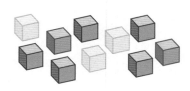

On this fair spinner, the probability of landing on a green section is $\frac{2}{5}$.

If you choose a cube at random, the probability of getting yellow is $\frac{3}{10}$.

If you pick a number card at random, the probability of getting a prime number is $\frac{4}{7}$.

When all possible outcomes are equally likely you can use this formula:

▶ Theoretical probability = $\dfrac{\text{number of outcomes favourable to the event}}{\text{total number of outcomes}}$

A **sample space** diagram is a useful way of listing all the possible outcomes.

example

The diagram shows all the possible outcomes when two dice are thrown and the numbers on the faces are added.
Find the probability of getting a score of 5.

....

There are 36 outcomes altogether.
There are 4 outcomes leading to a score of 5.

The probability of a 5 = $\dfrac{\text{number of outcomes leading to a 5}}{\text{total number of outcomes}}$

$$p(5) = \frac{4}{36}$$
$$= \frac{1}{9}$$

	2	3	4	5	6	7
	3	4	5	6	7	8
	4	5	6	7	8	9
	5	6	7	8	9	10
	6	7	8	9	10	11
	7	8	9	10	11	12

You can write p(5) to mean 'the probability of scoring 5'.
p(n) is the probability of any outcome n occurring.

Exercise D2.1

1 James has 36 blue and yellow counters. When he chooses a counter at random, it is equally likely that he will pick either a blue or a yellow counter. How many counters of each colour are there? Explain your answer.

2 Mandy puts three red cubes, two green cubes and five blue cubes into a bag. She asks Javid to pick a cube without looking. Javid says:

> There are three different colours, so the probability of the cube being green is $\frac{1}{3}$.

a Explain why Javid is wrong.
b Work out the correct probability of Javid picking a green cube.
c Mark the probability of picking each colour on a probability scale.

3 Draw accurate diagrams of circular spinners, like the one on page 138, which meet these conditions:
a A spinner with 10 equal sections; the probability of the pointer landing on a black section is 30%.
b A spinner with 12 equal sections coloured red, blue or green; there is an equal probability of the pointer landing on each colour.
c A spinner with 6 equal sections; the probability of getting black, yellow or orange is $\frac{1}{2}$, $\frac{1}{6}$ and $\frac{1}{3}$ respectively.

4 Tickets numbered 1 to 25 are placed in a bag. One ticket is chosen at random. Find the probability that the number chosen will be:
a an even number
b less than 6
c a prime number.
Write each of your answers as a fraction, a decimal and a percentage.

5 Ben has two bags of marbles. Bag A has 3 green and 2 red marbles. Bag B has 5 green and 4 red marbles. Ben is going to pick a marble at random from either bag A or bag B. Which bag should Ben choose if he wants a green marble? Explain your reasoning.

6 Karen has number cards marked 2, 4, 6. Jenna has cards marked 3, 5, 7. They each pick one of their cards at random, and the numbers on the two cards are multiplied together.
a Draw a sample space diagram to show all the possible answers.
b Use your diagram to find the probability that the answer will be less than 20.

7 A washing machine has four different washing temperatures: 30 °C, 40 °C, 60 °C and 90 °C. It has five different washing programs: A, B, C, D and E. One combination of temperature and washing program can be written as (30, A).
a Write down all of the combinations in the same way.
Jasmine selects a wash temperature and program at random. Her new jumper is in the wash, and will be safe provided that:
▶ the wash temperature is **lower** than 60 °C, and
▶ the washing program is **not** E.
On your list of combinations, circle the favourable outcomes for the event 'Jasmine's jumper gets washed safely'.
b Find the probability that Jasmine's jumper gets washed safely.

This spread will show you how to:
▶▶ Identify all the mutually exclusive outcomes of an experiment.
▶▶ Know that the sum of probabilities of all mutually exclusive outcomes is 1 and use this when solving problems.

KEYWORDS
Mutually exclusive
Experiment   Trial
Expected frequency

Mutually exclusive outcomes cannot happen at the same time.

When a coin is tossed there are two mutually exclusive outcomes: head or tail.

When a dice is rolled the outcomes '4' and 'an even number' are not mutually exclusive: 4 is even so they happen together.

▶ The sum of the probabilities of all mutually exclusive outcomes is 1.

example

A window in a fruit machine can show one of three different symbols.
The probabilities are shown in the table.
Find the missing probability.

Symbol	Probability
Star	0.35
Apple	0.25
Bell	

p(Star) + p(Apple) = 0.35 + 0.25 = 0.6
p(Bell) + 0.6 = 1,   so p(Bell) = 0.4

If you know the probability of an event, you can estimate how many times it is likely to occur.

Rolling a dice 50 times is an **experiment**.
Each roll of the dice is a **trial**.

▶ Expected frequency = probability × number of trials

example

A telephone company keeps records of calls made to mobile phones.

The table shows the probability of the different possible outcomes.

Result of call	Probability
Phone on, call answered	0.4
Phone on, but not answered	0.1
Phone on, but engaged	0.1
Phone off	

a Find the missing probability.
b Estimate the number of times you would expect a mobile phone to be switched off if 30 calls were made.

a 0.4 + 0.1 + 0.1 = 0.6
p(phone off) = 1 − 0.6 = 0.4
b 30 × 0.4 = 12 so you would expect the phone to be switched off 12 times.

p(phone off) is shorthand for 'the probability that the phone is off'.

Exercise D2.2

1 The table shows the probability of three players winning a competition.

Player	Geraint	Sally	Zia
Probability of winning	$\frac{1}{2}$	20%	0.3

Work out the probability of each player **not** winning.

2 Sam makes a spinner with three sections marked A, B and C.
Sam says: 'There is an evens chance of the pointer landing on A. The probability of getting B is 30%, and the probability of getting C is a quarter.'
Explain why Sam must be wrong.

3 The table shows the probability of four players winning a game.

Player	Karis	Fiona	Laura	Gill
p(winning)	20%	35%	30%	15%

Find the probability of each player **not** winning the game.

4 Grant works in an office at an engineering company. He uses a machine that is shared with another office. The table shows probabilities for the status of the machine.

Machine status	Probability
In office: being used	0.22
In office: not being used	
Out of office	0.4

a Find the probability that the machine is in the office, but not being used.
b The probability of another piece of equipment being out of the office is 0.37.
Find the probability that it is in the office.

5 Four different five-sided spinners are all marked with letters A, B, C, D and E.
The table shows the probability of getting each result for each spinner.

Result	A	B	C	D	E
Spinner 1	0.3	0.2	0.1	0.1	0.3
Spinner 2	$\frac{1}{2}$	0.2	20%	0.1	$\frac{1}{10}$
Spinner 3	28%	36%	27%	5%	5%
Spinner 4	$\frac{1}{4}$	$\frac{1}{8}$	$\frac{1}{4}$	$\frac{1}{8}$	$\frac{1}{4}$

The results for two of the spinners are wrong.
Find out which spinners have incorrect results. Explain your answers.

6 Jake takes a book from a shelf at random. The table shows the probability of getting different types of book.

Subject	Hardback	Paperback
Fiction	0.1	0.3
Sport	0.2	0
Computers	0.1	0.15
Animals	0.05	0.1

What is the probability that the book chosen will be:
a a paperback
b a book about computers.
There are 60 books on the shelf altogether.
c How many of them are books about animals?

7 There are four possible results in a fairground game. The table shows the probability of each result.

Result	Probability
Top prize	$\frac{1}{20}$
Consolation prize	$\frac{1}{10}$
Your money back	?
Lose	$\frac{3}{5}$

a What is the probability of getting your money back?
b What is the most likely result?
c What is the probability of not winning the top prize?
d What results would you expect if you played the game 100 times?

This spread will show you how to:
▶▶ Identify all the mutually exclusive outcomes of an experiment.
▶▶ Know that the sum of probabilities of all mutually exclusive
   outcomes is 1 and use this when solving problems.

KEYWORDS
Common denominator
Success   Random
Expected

You can write probabilities as fractions, decimals or percentages.

example

A letter from the word DAFFODIL is chosen at random.
What is the probability of choosing a vowel?
Give your answer as:
a a fraction   **b** a decimal   **c** a percentage.

....

a There are 3 vowels, and 8 letters in total
       p(vowel) = $\frac{3}{8}$
b 3 ÷ 8 = 0.375
c 0.375 × 100% = 37.5%

To add or subtract fractions, you use a **common denominator**.

example

A spinner has sectors coloured red, blue and green.
The probability of getting red is $\frac{1}{2}$, and the probability of getting
blue is $\frac{1}{3}$.
What is the probability of getting green?

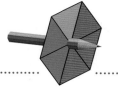

....

p(red or blue) = $\frac{1}{2} + \frac{1}{3}$

▶ First find a common denominator of 2 and 3:
   6 is a multiple of both 2 and 3.
▶ Convert to equivalent fractions: $\left(\frac{1}{2}\right) + \left(\frac{1}{3}\right)$

   $= \frac{3}{6} + \frac{2}{6}$

   The sum of all mutually
   exclusive outcomes is 1.

▶ Add the numerators: $= \frac{5}{6}$
   p(green) = 1 − p(red or blue) = $1 - \frac{5}{6} = \frac{6}{6} - \frac{5}{6} = \frac{1}{6}$

You can work out the expected number of successes in an experiment using the probability.

▶ **Expected number of successes = probability of success × number of trials**

example

The probability of winning a guessing game on a computer is $\frac{1}{4}$.
How many times would you expect to win if you played 60 times?

Fraction × integer
$= \frac{\text{numerator} × \text{integer}}{\text{denominator}}$

....

The expected number of wins is $\frac{1}{4} × 60 = 1 × \frac{60}{4} = \frac{60}{4} = 15$

Exercise D2.3

1 A bag contains these numbers of coloured counters.

Colour	Pink	Blue	Yellow	Green	Orange
Number	4	6	9	12	5

A counter is chosen at random from the bag. Find the probability of the counter being:

a pink   **b** blue
c yellow   **d** green
e orange   **f** not blue.

Give each answer as:

i a fraction in its lowest terms
ii a percentage.

2 A bag of mixed sweets contains four different flavours.

Flavour	Lemon	Orange	Sherbet	Mint
Number of sweets	11	12	8	9

Alan likes all the flavours except lemon.
Beth only likes orange and mint.
Claire likes everything except orange.
Each person chooses a sweet from the bag at random.

a Find the probability of each person getting a sweet they like.
Give your answers as fractions in their lowest terms.
b Change each answer to a percentage.
c Mark each of the probabilities on a probability scale.

3 The probability of getting a winning ticket in six different lotteries is shown in the table.
Copy and complete the table.

Lottery	1	2	3	4	5	6
Probability of winning	$\frac{1}{2}$	$\frac{1}{3}$	$\frac{2}{5}$	$\frac{3}{8}$	$\frac{4}{17}$	$\frac{9}{125}$
Probability of not winning						

4 Roger picks a coloured cube at random from a bag.
The probability of getting a blue cube is $\frac{3}{25}$.
The probability of getting a yellow cube is $\frac{8}{25}$.
What is the probability of:
a **not** getting a blue cube
b getting a blue **or** a yellow cube?

5 When two ordinary dice are rolled and the scores are added, the probability of a total of 7 is $\frac{1}{6}$ and the probability of a total of 2 is $\frac{1}{36}$.

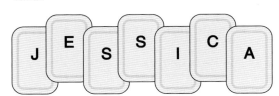

What is the probability of:
a not getting a total of 7
b not getting a total of 2
c getting a total of 2 **or** 7?

6 Jessica writes each letter of her name on a card.

J E S S I C A

She chooses a card at random, notes the letter, and then puts it back in the pack.
a How many times would you expect her to get a letter S if she repeated the experiment 50 times?
b How many vowels would you expect Jessica to get in 40 trials?
c How many times would you expect her to get a J or a C in 60 trials?

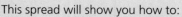
This spread will show you how to:

▶▶ Estimate probabilities from experimental data.

KEYWORDS
Experimental probability
Trial   Successful
Bias   Data

You can estimate probabilities by carrying out an experiment.

▶ Experimental probability = $\dfrac{\text{number of successful trials}}{\text{total number of trials}}$

example

Jenny is dropping counters onto a chessboard.
She wants to know the probability that a counter will land on a line.
Jenny drops 20 counters and counts how many touch a line.
She repeats this ten times and gets these results:

| 10 | 16 | 14 | 12 | 15 | 13 | 13 | 14 | 15 | 18 |

Estimate the probability that a counter will land on a line.

....

$20 \times 10 = 200 \implies$ there are 200 trials altogether
$10 + 16 + 14 + 12 + 15 + 13 + 13 + 14 + 15 + 18 = 140$
$\implies$ 140 trials are 'successful'
Experimental probability $= \dfrac{140}{200} = 0.7$

▶ The more trials you carry out, the more reliable your estimate of the experimental probability will be.

Dice, counters and spinners are examples of apparatus.

You can use experimental probability to decide whether a piece of apparatus is fair or **biased**.

example

Four students tested a dice to see if it was fair or biased. The table shows their results.
Find the:
a experimental probability of getting a 6
b theoretical probability of getting a 6.

Student	Number of trials	Number of sixes
Barry	100	27
Simon	150	35
Davina	75	21
Hannah	200	51

....

a Combine all the results:
$27 + 35 + 21 + 51 = 134$ (sixes) $100 + 150 + 75 + 200 = 525$ (trials)
Experimental probability $= \frac{134}{525} = 0.26$ (2 dp)
b Theoretical probability $= \frac{1}{6}$ (six equally likely outcomes)
$= 0.17$ (2 dp)
0.26 is bigger than 0.17, so the dice is probably biased towards sixes.

▶ A piece of apparatus is biased if the experimental probabilities, based on a large number of trials, are very different from the theoretical probabilities.

Exercise D2.4

1 State whether these statements are true or false. Explain your answers.

  a If you roll a fair dice, the probability of getting a 6 is $\frac{1}{6}$.

  b If you roll a fair dice six times, you are certain to get a 6.

  c If you get three heads in a row when you toss a fair coin, you are more likely to get a tail next time.

2 Sam tests a dice to see whether it is fair. The table shows his results.

Score	1	2	3	4	5	6
Frequency	17	22	13	14	23	31

  a Estimate the probability of each score.

  b Do you think that the dice is fair? Explain your answer.

Sam's friends help him collect more results:

Score	1	2	3	4	5	6
Frequency	170	159	164	148	185	174

  c Estimate the probability of each score, based on the complete set of results.

  d Do you now think that the dice is fair? Explain your reasoning.

3 The table shows the number of children living in the first 20 houses in a street.

Number of children	Number of houses
0	5
1	6
2	7
3	2

  a Estimate the probability that the number of children living in a randomly chosen house in the street is at least two.

  b How reliable would your answer to part **a** be if there were:

    i 25 houses in the street altogether

    ii 200 houses altogether?

    Explain your reasoning.

4 A spinner is tested by four students to see if it is biased. Their results are shown in the table.

Student	Number of trials	Colour				
		Red	Blue	Green	Yellow	Black
Jolomi	200	43	41	22	19	75
Theo	25	8	7	0	2	8
Carol	100	24	22	18	10	26
Alice	75	15	23	6	8	23
Totals						

  a Which student's data should give the best estimates for experimental probability? Explain your answer.

  b The four students decided to collect all of their data together.
Copy the table, and fill in the totals.

  c From the evidence in the table, do you think that the spinner is biased? Explain your answer.

  d Use the data to estimate the probability of the spinner landing on black.

5 Mike puts 25 coloured cubes in a bag. He asks five friends to estimate how many cubes of each colour there are.
Each person picks a cube at random, records the colour, and puts it back in the bag.
They repeat this 200 times. Here are their results:

Name	Black	White	Pink	Brown
Patrick	48	83	49	20
Sandra	36	86	53	25
Claire	29	88	55	28
Jon	43	76	59	22
Maria	40	83	59	18

  a Use Patrick's data to estimate the number of cubes of each colour.

  b Combine all the students' data and use this to work out another estimate of the number of cubes of each colour.

  c Which result should be more reliable, part **a** or part **b**? Explain your answer.

You should know how to …

1 Know that the sum of probabilities of all mutually exclusive outcomes is 1 and use this when solving problems.

2 Present a concise, reasoned argument using diagrams and related explanatory text.

Check out

1 A jar contains only red, green and orange sweets.

When a sweet is chosen at random, the probability of getting a red sweet is 0.2, and the probability of getting a green sweet is 0.35.

Find the probability of getting an orange sweet. Show your working carefully.

2 There are two dice games on a stall at a school fair.

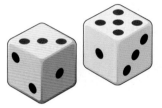

It costs 10p to play either game, and you get 50p if you win. The rules are:

Game A
Roll two ordinary dice and add the numbers together.
You win if the total is greater than 10.

Game B
Roll two ordinary dice and multiply the numbers together.
You win if the product is greater than 20.

Which game would you rather play?
Explain your reasoning carefully, using appropriate vocabulary and diagrams.

Transformations and congruence

This unit will show you how to:

- ▶▶ Transform 2-D shapes by combinations of translations, rotations and reflections.
- ▶▶ Know that translations, rotations and reflections map objects onto congruent images.
- ▶▶ Identify reflection symmetry in 3-D shapes.
- ▶▶ Enlarge 2-D shapes, given a centre of enlargement and a whole-number scale factor.
- ▶▶ Identify the scale factor of an enlargement as the ratio of lengths of line segments.
- ▶▶ Recognise that enlargements preserve angle but not length, and understand the implications for perimeter.
- ▶▶ Use and interpret maps and scale drawings.
- ▶▶ Use proportional reasoning to solve a problem.
- ▶▶ Represent problems and synthesise information in geometric form.
- ▶▶ Present a concise, reasoned argument.

Shapes can be transformed to make attractive patterns.

Before you start

You should know how to ...

1 Reflect, rotate, translate and enlarge a shape.

2 Identify 3-D shapes.

3 Calculate with ratios, including mixed units.

Check in

1 Copy the diagram. Plot the image of △ABC after these transformations:

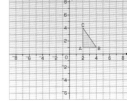

a Reflect in the y-axis.

b Rotate 90° clockwise about the origin.

c Translate by $\begin{pmatrix} 3 \\ -4 \end{pmatrix}$.

2 Name these shapes:

a   b   c

3 Reduce these ratios to their simplest terms:

   a 5 : 20   b 14 : 35   c 25 : 40   d 24 : 92

This spread will show you how to:
- ▶▶ Understand and use the language and notation associated with reflections, translations and reflections.
- ▶▶ Recognise and visualise the transformation of a 2-D shape.

KEYWORDS

Reflection	Vector
Rotation	Convention
Centre of rotation	
Translation	Mirror line
Direction	Image

You should know how to reflect, rotate and translate a shape.

△ABC has been reflected in the mirror line to give the image A'B'C'.

△ABC has been translated by $\begin{pmatrix} 4 \\ -2 \end{pmatrix}$ to give the image A'B'C'.

△ABC has been rotated by 90° clockwise about X to give the image A'B'C'.

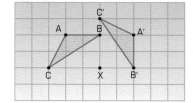

▶ You specify a ...
  - ▶ reflection by a **mirror line**
  - ▶ translation by a **vector**
  - ▶ rotation by a **direction**, an **angle** and a **centre of rotation**.

By convention, anticlockwise is taken as the positive direction of rotation.

example

The diagram shows a parallelogram ABCD on a grid.
a **i** Reflect ABCD in the mirror line M.
Label the image A'B'C'D'.
ii What is the equation of the mirror line?
b Translate ABCD by the vector $\begin{pmatrix} -3 \\ -2 \end{pmatrix}$.
Label the image A″B″C″D″.
c Rotate ABCD through 90° anticlockwise about the origin.
Label the image A‴B‴C‴D‴.

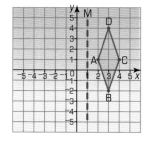

a **i** 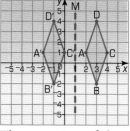 **b** 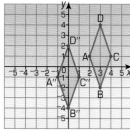 **c**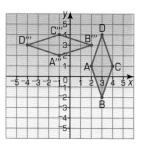

ii The equation of the mirror line is $x = 1$.

Exercise S3.1

1 Make five copies of this diagram.

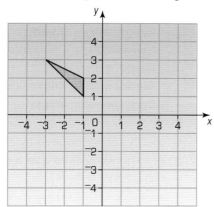

Using a new copy of the diagram each time, reflect the triangle in each of these mirror lines.

a $x = {}^-1$ **b** $y = 2$
c $x = 1$ **d** the x-axis
e the y-axis.

2 Make six copies of this diagram.

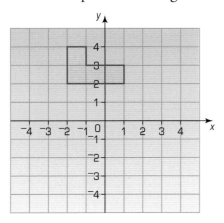

Using a new copy of the diagram each time, draw the image after each of these rotations.

a 90° clockwise about (1, 2)
b $\frac{1}{2}$ turn about (1, 3)
c $\frac{1}{2}$ turn about (0, 0)
d 90° clockwise about (0, 0)
e 90° anticlockwise about ($^-2$, 2)
f $\frac{1}{2}$ turn about ($^-1$, $^-2$).

3 Make three copies of this diagram.

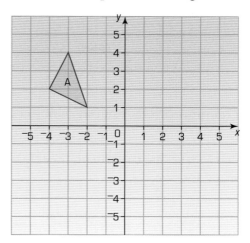

On the first copy:
a Reflect shape A in the x-axis.
   Label the reflection B.
b Reflect A in the y-axis.
   Label the reflection C.

On the second copy:
c Rotate A by $\frac{1}{4}$ turn anticlockwise, centre (0, 0).
   Label the rotation P.
d Rotate A by $\frac{1}{4}$ turn clockwise, centre (0, 0).
   Label the rotation Q.
e Rotate A by $\frac{1}{2}$ turn, centre (1, 0).
   Label the rotation R.

On the third copy:

f Translate A using the vector $\begin{pmatrix} 5 \\ -2 \end{pmatrix}$.

   Label the translation X.

g Translate X using the vector $\begin{pmatrix} -2 \\ 6 \end{pmatrix}$.

   Label the translation Y.

h What translation will move Y back to A?

Describing transformations

This spread will show you how to:
- ▶▶ Understand congruence.
- ▶▶ Know that translations, rotations and reflections preserve length and angle and map objects onto congruent images.

KEYWORDS
Congruent   Object
Convention
Image
Corresponding sides
Corresponding angles

Many transformations result in an object and image that are identical.

If two 2-D shapes are **congruent**, they have the same shape and size.

This means that corresponding sides and angles will be equal.

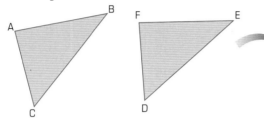

Triangles ABC and DEF are congruent.

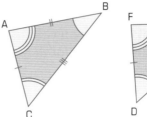

Show corresponding lengths and angles using the standard conventions.

▶ **You can map congruent shapes onto each other by a reflection, a rotation or a translation.**

example

The shapes A to E are congruent.
Describe a transformation that maps the object A onto each of the image shapes B to E.

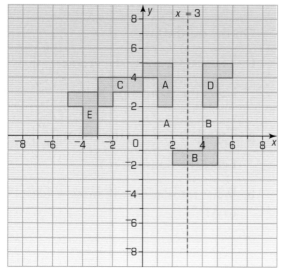

A → B: Rotation of ⁻90°, centre (0, 0)

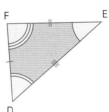

A → C: Rotation of ⁺90°, centre (0, 2)

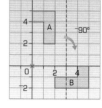

A → D: Reflection in the line $x = 3$

A → E: Translation by $\begin{pmatrix} -5 \\ -2 \end{pmatrix}$

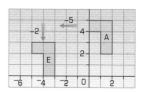

Exercise S3.2

1 Which of these shapes are congruent?

2 The diagram shows six congruent triangles
labelled A to F.
Describe these transformations:
  a A to B
  b B to C
  c E to F
  d C to E
  e C to B
  f B to A.

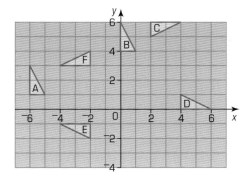

3 The diagram shows six congruent shapes
labelled A to G.

A is the object.

Match each of these transformations
with the correct image.
  a reflection in $x = {}^-1$
  b rotation 90° clockwise about (0, 0)
  c reflection in $y = 0$
  d rotation 180° anticlockwise about
     (0, 0)
  e reflection in $y = 3$
  f translation with vector $\begin{pmatrix} 1 \\ -3 \end{pmatrix}$.

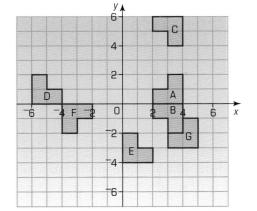

4 Copy the diagram and draw these
transformations of shape A:
  a reflection in $y = 2$ (label B)
  b reflection in $x = 2$ (label C)
  c 180° rotation about (0, 0) (label D)
  d translation with vector $\begin{pmatrix} 8 \\ -4 \end{pmatrix}$ (label E)
  e rotation 90° clockwise, centre (0, 2)
    (label F).

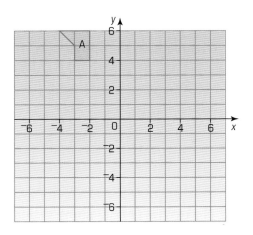

Combining transformations

This spread will show you how to:
▶▶ Transform 2-D shapes.
▶▶ Know that translations, rotations and reflections map objects onto congruent images.

KEYWORDS
Image   Transformation
Map

You can often write a combination of transformations using a single transformation.

example

a Reflect shape A in the line $x = 1$. Label the image A′.
b Reflect shape A′ in the line $y = {}^-1$. Label the image A″.
c What single transformation will map A onto A″?
d Reflect shape A′ in the line $x = 3$. Label the image A‴.
e What single transformation will map A onto A‴?

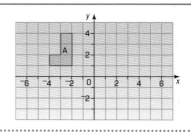

a b and **d**

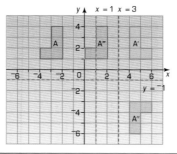

c A ⟶ A″ is a rotation of 180°, centre $(1, {}^-1)$.

e A ⟶ A‴ is a translation by $\begin{pmatrix} 4 \\ 0 \end{pmatrix}$.

The order of transformations can make a difference to the result.

example

ABC is a triangle on a grid.
Transformation X is a reflection in the x-axis.
Transformation Y is a rotation of 90° clockwise, centre $(0, 0)$.

a Map △ABC onto image A′B′C′ by applying X then Y.
b Map △ABC onto image A″B″C″ by applying Y then X.
c Comment on the result.

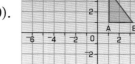

a

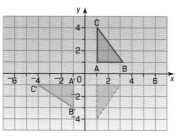

b

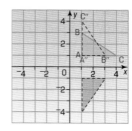

c The two images are not the same.

Exercise S3.3

1 Copy the diagram.

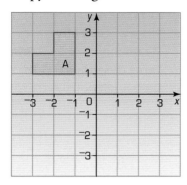

a Reflect shape A in the *y*-axis and label the new shape A′.
b Reflect A′ in the *x*-axis and label it A″.
c What single transformation will transform A to A″?

2 Repeat question 1 but change the order of the transformations:

► Reflect shape A in the *x*-axis and label the new shape A′.
► Reflect A′ in the *y*-axis and label it A″.

What do you notice?

3 Copy the diagram.

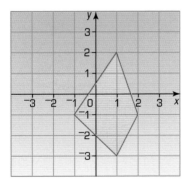

a Reflect the quadrilateral in the *x*-axis.
b Use colour to show which lines and angles are equal in size.

4 Copy the diagram.

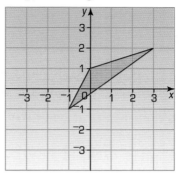

a Reflect the triangle in the *x*-axis.
b Reflect both shapes in the *y*-axis.
c Use colour to show which lines and angles are equal in size.

5 Copy the diagram.

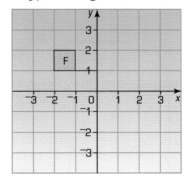

a Rotate the flag F through 90° clockwise centre (0, 0) and label the new flag F′.
b Reflect F′ in the *x*-axis and label it F″.
c Show that F can be transformed to F″ by reflection in $y = x$.
d Describe other ways of transforming F to F″.

6 Investigate combinations of reflections, for example the effect of a reflection in $x = 2$ followed by a reflection in $y = 2$.

7 Investigate combinations of rotations. What single transformation is equivalent to two rotations?

S3.4 Planes of symmetry

This spread will show you how to:
▶▶ Identify reflection symmetry in 3-D shapes.

KEYWORDS
Plane of symmetry
Pyramid   Prism
Three dimensional (3-D)
Cross-section

A **cross-section** is a plane through a 3-D shape.
It is a plane of symmetry if the two parts are identical.

This cuboid has three planes of symmetry.

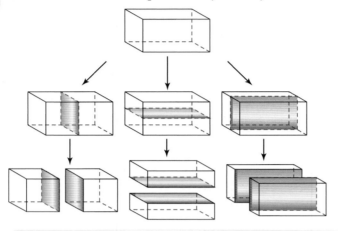

▶ A **plane of symmetry** divides a shape into two identical
halves so that each half is a mirror image of the other.

A cuboid is a type of **prism**.
It has a uniform cross-section
throughout its length.
All prisms have at least one
plane of symmetry.

The diagram shows a square-based pyramid.

a Shade in a horizontal cross-section through the pyramid.
Describe its shape. Is the cross-section a plane of symmetry?
b Shade in all the planes of symmetry through the pyramid.

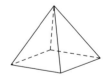

....

a
The cross-section is a square.
It is not a plane of symmetry.

b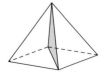

The square-based pyramid has four planes of symmetry.

Exercise S3.4

1 Copy these shapes and draw in any lines of symmetry.

a

b

c

d

2 Make sketches to show the planes of symmetry of these shapes:

a cube

b triangular prism

c regular tetrahedron

d regular octahedron.

3 There are many ways to slice a cube.

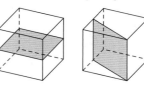

Can you slice a cube so that the cross-section is:

a a square
b a rectangle
c a triangle
d a circle
e a pentagon
f a hexagon?

Give reasons for your answers.

4 Repeat question 3 using a cuboid rather than a cube.

5 Investigate other 2-D shapes you can make when you slice other 3-D shapes.

6 You can slice a cone in many different ways:

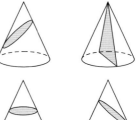

Explain how to slice a cone so that the cross-section is:

a a circle
b an isosceles triangle
c an oval.

This spread will show you how to:

▶▶ Enlarge 2-D shapes.

▶▶ Identify the scale factor of an enlargement.

▶▶ Recognise that enlargements preserve angle but not length.

KEYWORDS

Enlargement   Proportion
Centre of enlargement   Ratio
Scale factor   Similar
Corresponding sides

In an enlargement:

▶ the angles are the same

▶ the lengths increase in the same proportion.

example

Enlarge the trapezium ABCD by scale factor 2 with centre of enlargement (1, 2).

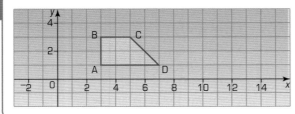

1 Draw lines from (1, 2) to each of the vertices.

2 Extend the lines to twice their original length.

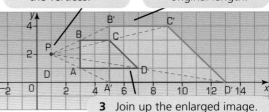

3 Join up the enlarged image.

▶ To specify an enlargement, you need a **centre of enlargement** and a **scale factor**.

example

Rectangle A'B'C'D' is an enlargement of ABCD.
Identify the scale factor and the centre of enlargement.

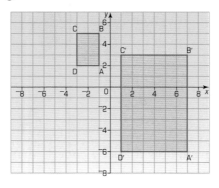

2 Extend the lines to the centre of enlargement, (⁻5, 6).

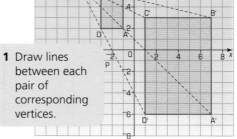

1 Draw lines between each pair of corresponding vertices.

To find the scale factor, take any two corresponding line segments:
A'B' = 9   AB = 3
The ratio A'B' : AB = 9 : 3 = 3 : 1
The scale factor is 3.

▶ The ratio of corresponding line segments is the scale factor.

Exercise S3.5

1 Write down the dimensions of this book following an enlargement of 10%.
[Its size is 246 mm × 189 mm.]

2 Copy these diagrams onto squared paper and draw enlargements with the given scale factor and centre, C.

a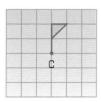

Scale factor 3

b

Scale factor 2

c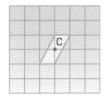

Scale factor 3

d

Scale factor 3

3 Copy this grid four times.

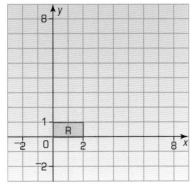

Using a new grid each time, draw each of these enlargements of shape R:
a scale factor of 2, centre (0, 0)
b scale factor of 2, centre (1, 1)
c scale factor of 3, centre (⁻1, 0)
d scale factor of 2, centre (⁻2, ⁻2).

4 Find the scale factor and the coordinates of the centre of enlargement.
The small shape in each case is the object.

a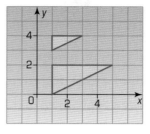

b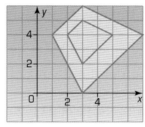

5 Investigation
a Sketch three triangles which are similar to this one.

> Enlargements are similar shapes.

b Find the perimeter of each triangle in part **a**.
What do you notice?
c Draw some other simple shapes and enlarge them to different scale factors. Investigate the relationship between enlargement and perimeter.

6 A shape has an area of 12 cm² and a perimeter of 14 cm.
Find the area and perimeter of the shape following an enlargement with scale factor 3.

S3.6 Scale and proportion

This spread will show you how to:
- ▶▶ Use and interpret maps and scale drawings.
- ▶▶ Use proportional reasoning to solve a problem.

KEYWORDS

Map   Proportional
Scale   Scale drawing
Unitary method

You can show distances on a **map**.

This map is drawn to a scale of 1 cm to 2 km.

The distance between Alphatown and Betaville on the map is 4.7 cm.

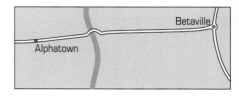

The **real** distance between the two towns is given by:

| 1 cm | $\rightarrow$ | 2 km |
| 4.7 cm | $\rightarrow$ | 4.7×2 km = 9.4 km |

You can use proportional reasoning to solve map problems.

---

example

A map is drawn to a scale of 2 cm to 5 km.
The real distance between two villages is 18 km.
Find the distance between the towns on the map.

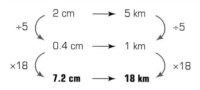

This is an example of the **unitary** method.
You find out the value of **one** unit, then scale it up or down.

So the two towns are 7.2 cm apart on the map.

You also use scale in a scale drawing.

example

Daniel is making a scale drawing of his bedroom.
Daniel's bedroom has a length of 3 m.
The scale drawing has a width of 3 cm and a length of 4.5 cm.

a Find the scale of the drawing.
b Find the width of Daniel's bedroom.

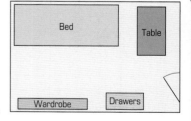

a Compare the lengths:

| | 4.5 cm | $\rightarrow$ | 3 m | |
| | 1.5 cm | $\rightarrow$ | 1 m | The scale is 1.5 cm to 1 m. |

b

| | 1.5 cm | $\rightarrow$ | 1 m | |
| So | 3 cm | $\rightarrow$ | 2 m | Daniel's bedroom has a width of 2 m. |

You can write a scale as a ratio:

    1.5 cm : 1 m
= 1.5 cm : 100 cm
=   3 : 200

Use whole numbers in ratios.

Exercise S3.6

1 Here is a map of Luppitt village.
The scale is 4 cm to 1 km.

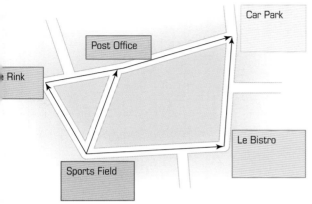

a How far is it from the Sports Field to the Ice Rink?

b How far is it from the Post Office to the Car Park?

c Sarah walks from the Post Office to the Bistro by the shortest route. How far has she walked?

d A new map is made using a scale of 2 cm to 1 km. On the map, what will be the distance from the Sports Field to the Ice Rink?

2 The scale on a map is 1 cm to 2 km.

a If the distance between York and Scarborough is 60 km, find the distance between the towns on the map.

b On the map, the distance between Scalby and Ravenscar is 5.5 cm. Find the actual distance.

c Write the scale in the form 1 : *n*.

The scale of a map is 2 cm to 5 km.

d If the distance between Poole and Plymouth is 100 km, find the distance between the towns on the map.

e On the map, the distance between Plymouth and Taunton is 8.5 cm. Find the actual distance.

f Write the scale in the form 1 : *n*.

For questions 3 to 8, look at this plan of a garden. It is drawn to a scale of 0.5 cm to represent 2 m.

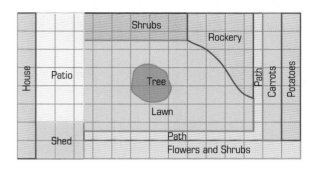

3 How long and wide is the garden?

4 Find the area of the garden, and the area of the lawn in m^2.

5 How wide is the path in cm?

6 Paving slabs come in three sizes:
2 feet × 3 feet, 2 feet × 2 feet, and 2 feet × 1 foot.
(1 foot is about 30 cm)

a Which slabs would be best to use to cover the patio?

b How many slabs are needed?

7 When the tree is fully grown the diameter of the branches will be about 3 m.
This means that the tree should be at least 5 m from the house.
Is the tree planted too near the house?

8 Carrots should be planted in rows 20 cm apart.
How many rows of carrots can be planted longways in the carrot bed?

Summary

You should know how to ...

1 Know that translations, rotations and reflections map objects onto congruent images.

2 Use proportional reasoning to solve a problem.

3 Present a concise, reasoned argument.

Check out

1

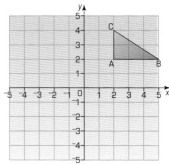

Copy the diagram.

a Rotate △ABC 90° anticlockwise about centre (0, 0). Label your image A′B′C′.

b Reflect A′B′C′ in the x-axis. Label your image A″B″C″.

c What single transformation will transform ABC onto A″B″C″?

2

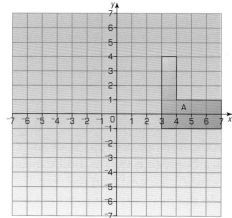

Copy the diagram.

a Reflect shape A in the line y = ⁻1 and label the image B.

b Reflect A in the y-axis and label the image C.

c Reflect B in the y-axis and label the image D.

d Reflect C in the line y = ⁻1 and label the image E.

e What do you notice?

3 Can you have a hexagon with four lines of symmetry?
Explain and justify your answer.

Problem solving and revision

This unit will show you how to:

- ▶▶ Solve increasingly demanding problems and evaluate solutions.
- ▶▶ Explore connections in mathematics.
- ▶▶ Represent problems and synthesise information in different forms.
- ▶▶ Solve substantial problems by breaking them down into simpler steps and tasks.
- ▶▶ Use trial and improvement.
- ▶▶ Present a concise, reasoned argument.

- ▶▶ Identify exceptional cases or counter-arguments.
- ▶▶ Give solutions to an appropriate degree of accuracy.
- ▶▶ Use proportional reasoning to solve a problem.
- ▶▶ Interpret and use ratio in a range of contexts.
- ▶▶ Solve problems using properties of angles and polygons, justifying inferences and explaining reasoning.
- ▶▶ Check results using appropriate methods.

You will use four stages in an exam question to maximise your marks:

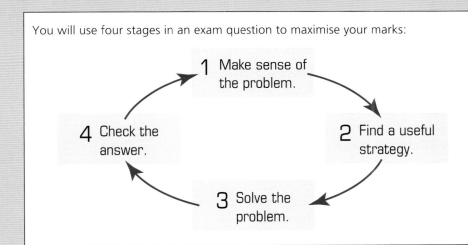

1 Make sense of the problem.

2 Find a useful strategy.

3 Solve the problem.

4 Check the answer.

Before you start

You should know how to ...

- ▶ Calculate with integers, fractions, decimals and percentages.
- ▶ Use proportional reasoning and ratio.
- ▶ Make and justify estimates and approximations.
- ▶ Use algebra to represent unknown values.
- ▶ Construct and solve linear equations.
- ▶ Use formulae.
- ▶ Write an expression for the general term of a sequence.
- ▶ Use angle facts and properties of polygons.

- ▶ Visualise 2-D representations of 3-D shapes.
- ▶ Transform 2-D shapes.
- ▶ Construct 2-D shapes.
- ▶ Use units of measurement.
- ▶ Know and use formulae for perimeter, area and volume.
- ▶ Find summary values for data.
- ▶ Interpret graphs and diagrams.
- ▶ Use probabilities.

This spread will show you how to:
▶▶ Solve increasingly demanding problems.
▶▶ Explore connections across mathematics.
▶▶ Represent problems and synthesise information.

The first stage is to understand what a question is asking.

Read the problem, and check that you know exactly what you need to find out, such as:
▶ a distance
▶ a weight
▶ an explanation
and so on.
▶ an area
▶ the name of something

Exam questions include a space for you to write your answer. The **units** may be written here. They will help you identify what you are being asked to find out.

example

Mike and Shahid are using their bicycles to measure distances.

a The front wheel on Mike's bicycle has a diameter of 70 cm.
Mike pedals his cycle so that his wheel goes round exactly 10 times.
Calculate how far Mike has moved.

b Shahid's front wheel has a diameter of 64 cm.
Shahid pedals his cycle exactly 1000 cm.
Calculate how many times the front wheel goes round.
Give your answer to two decimal places.

....

a You want to find the **distance travelled**.
Your answer will be in **centimetre**s.

Diameter = 70 cm
Circumference = πd = 3.14 × 70 cm
                              = 219.8 cm
The wheel turns around 10 times.
Distance travelled   = 10 × 219.8 cm
                              = 2198 cm

Write down the facts you know.

The formula is given on the exam paper.

b You want to find the **number of times the wheel turns round**.
Your answer will be a **number** (of times).

Diameter = 64 cm
Circumference = πd = 3.14 × 64
                              = 200.96 cm
Distance travelled   = 1000 cm
Number of turns   = 1000 cm ÷ 200.96 cm
                              = 4.98 times (2 dp)

Check the accuracy required.

Exercise P1.1

For each of these questions, write down what the question is asking you to find out before you solve the problem.

1   This is a pattern made with matchsticks:

   a Joanne wants to make the pattern with
      10 squares in it.
      How many matchsticks will she need?
   b Richard uses 51 matchsticks to make a
      pattern.
      How many squares are there in the
      pattern?

2   The graph shows the cost of buying and
   selling cappuccinos in Del's Cafe:

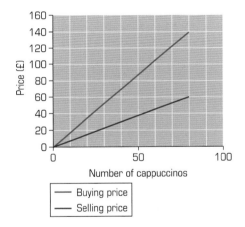

   a How much does Del pay for
      40 cappuccinos?
   b For how much does he sell
      55 cappuccinos?
   c Del spends £45 on cappuccinos and
      sells all of them.
      How much profit does he make?

3   Alan buys 6 trees for £110.70.
   How much would 11 trees cost?

4   Calculate:
   a 9% of £28.50
   b 17.5% of £212.

5   The shaded face of this prism is in the
   shape of a trapezium.
   Calculate the volume of the prism.

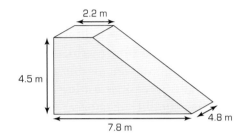

6   A large pond has a diameter of 4.8 metres.

   a Jonathon wants to put curved paving
      stones around the circumference of the
      pond.
      Each curved paving stone would cover
      35 cm of the circumference.
      Jonathon has 45 curved paving stones.
      Can Jonathon completely surround
      the pond with his paving stones?
   b Kevin wants to build a circular
      fountain with a radius of 1.4 m in the
      centre of the pond.

      What area of pond will be left?

Answering the question

This spread will show you how to:
- ▶▶ Solve demanding problems and evaluate solutions.
- ▶▶ Represent problems and synthesise information.
- ▶▶ Present a concise, reasoned argument.

Always check that you answer the whole question.
You will often be asked to explain your answers.

Highlight key words to help identify the information you need.

example

Ronnie has two bags of marbles, labelled X and Y.
Bag X contains **12 blue** marbles and **21 red** marbles.
Bag Y contains **10 blue** marbles and **18 red** marbles.
Ronnie will choose one marble at **random** from either bag.
Ronnie wants to **choose a blue marble**.
Which bag should she choose?
Show your working and **justify your answer**.

This question is about probability.

You want to know **which bag to choose**.
Your answer will be **X** or **Y**.
You must **explain** your choice.

Calculate the probability of choosing a blue marble for each bag, and see which bag has the greater probability.

	Bag X	Bag Y
Number of blue marbles	12	10
Number of red marbles	21	18
Total number of marbles	33	28

Using a table helps you keep track of information.

Bag X: Probability of a blue marble $= \frac{12}{33}$
Bag Y: Probability of a blue marble $= \frac{10}{28}$

You need to see which of the fractions is larger.
Using equivalent fractions ...

$$\frac{12}{33} = \frac{4}{11} = \frac{56}{154}$$
$$\frac{10}{28} = \frac{5}{14} = \frac{55}{154}$$
$$\frac{56}{154} > \frac{55}{154}$$
so $\quad \frac{12}{33} > \frac{10}{28}$

Always show your working so you can pick up marks even if you make a mistake.

So Ronnie has a better chance of choosing a blue marble with **Bag X** because $\frac{12}{33} > \frac{10}{28}$.

Exercise P1.2

1 These charts show information about the pets kept by pupils at Smallton School and Big City Academy.

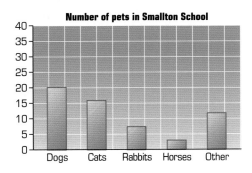

Number of pets in Smallton School

Number of pets at Big City Academy

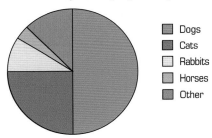

Dogs
Cats
Rabbits
Horses
Other

a Steve says: 'The charts show that there are more pet cats at Smallton School than at Big City Academy'.
Decide if you agree or disagree with Steve.
Use the charts to explain your answer carefully.

b Susan says: 'The charts show that a greater proportion of the pupils at Smallton School have rabbits as pets compared with Big City Academy'.

Decide if you agree or disagree with Susan.
Use the charts to explain your answer carefully.

2 **a** Which colour is this spinner most likely to land on?
Explain and justify your answer.

b Describe how to shade the spinner so that the chance of getting a red section is a third of the chance of getting a blue section.

3 Year 10 are trying to decide the colour of the new Year 11 school jumper. There are four colours to choose from – black, red, blue or pink.
All the pupils and teachers get one vote for their favourite colour. All the votes are then put in a box.
A voting card is then chosen at random from the box. Jade says that there are only four colours to choose from so the probability of the voting card being for the colour pink is $\frac{1}{4}$.
Explain why Jade is wrong.

4 This table shows some information about planets in the Coughalot galaxy.

Planet	Surface area (km²)	Population
Ticho	5.44 million	57.5 million
K'aargghhh	0.42 million	15.3 million
Hmmm 2	5.05 million	49.8 million
Ear Ak	2.44 million	32.6 million

a Which planet gives you the most room in which to live?
Explain and justify your answer.

b On the planet Ticho, all the land is shared out equally amongst the people who live there. How much land, in m^2, does each person own?

1 km² = 1 000 000 m²

165

Choosing a strategy

This spread will show you how to:

▶▶ Solve substantial problems by breaking them down.

▶▶ Use trial and improvement.

KEYWORDS

Perimeter   Equation

Trial and improvement

Once you understand the problem, you may need a strategy to help solve it.
Some problems need to be broken down into steps.

example

The edge of a square paving stone is *k* **centimetres**.
Paula arranges 5 paving stones like this:
The **perimeter is exactly 960 cm**.
What is the **length of a square stone**?

←*k*→

You want to know the **length of a stone**.
Your answer will be in **centimetres**.

You know that:

▶ Length of 1 stone = k

▶ Perimeter of this stone arrangement = 960 cm

Write down the facts you know
and the facts you need to
know.

You want to know the value of k.
Break the question down into parts:

1 Find an expression for the perimeter using algebra.
Perimeter = 12 × tile length = $12k$

2 Make an equation involving the perimeter.
Perimeter = $12k$ = 960 cm

3 Solve the equation to find the length of the square.
$$12k = 960 \text{ cm}$$
$$k = (960 \div 12) \text{ cm}$$
Length of stone = k = 80 cm

Another way of tackling a problem is by trial and improvement.

example

Two numbers **multiply to make** ⁻20 and **add to make 8**.
What are the two numbers?

You want to find the **value of two numbers**.
Your two answers will be **numbers**.

Call your two unknown numbers a and b.
You know that: $a + b = 8$ and $a \times b = {}^-20$

Choose a number for a, then b added to a makes 8.
Check if $a \times b = {}^-20$.

The two numbers are 10 and ⁻2.

a	b	a × b
5	3	15 too high
7	1	7 too high
9	⁻1	⁻9 too high
10	⁻2	⁻20

Exercise P1.3

1 From this list of numbers:

$$^-6 \quad ^-5 \quad ^-3 \quad ^-1 \quad 0 \quad 2 \quad 5 \quad 6 \quad 8$$

  a Find two numbers with a total of 1.
  b Find two numbers with a difference of 7.
  c Choose three numbers that have the lowest possible total.

2 Lindsey and Robert want to pack 44 bottles of pop into a box.
Each bottle weighs 570 g.
The box weighs 5 kg.
They read a notice that recommends the maximum weight for filling the box.

25 kg max

This box will not hold weights above 25 kg.

Is the box safe to lift?
Explain and justify your answer.

3 David has made up a number puzzle:

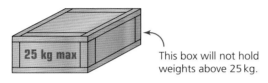

Multiplying my number by 3 and subtracting 4 gives the same answer as multiplying my number by 2 and adding 1.

Solve David's number puzzle.

4 Geoff and his father have a combined age of 62 years. In 5 years' time, Geoff will be exactly one third of his father's age.
How old is Geoff now?

5 Here are five expressions:

$$x - 3 \qquad 4x \qquad x^2 \qquad \frac{x}{5} \qquad \frac{10}{x}$$

  a Which expression gives the greatest value when x is between 0 and 1?
  b Which expression gives the smallest value when x is between 3 and 4?
  c Calculate the values of x that make each of the expressions equal to 1.

6 Cassie and Lucy each have four cards, numbered 2, 3, 4 and 5.

They each choose one of their own cards and multiply the numbers on the two cards together to make the answer.
  a Calculate the probability that the answer is an odd number.
  b The probability of the answer being less than x is $\frac{5}{8}$.
    Give a possible value for x.

7 This trapezium has an area of 20 cm^2.

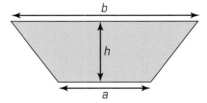

Copy and complete this table to show three possible values for h, b and a.

a	b	h
3 cm		
		2.5 cm

Proportional reasoning

This spread will show you how to:
▶▶ Solve problems involving percentage changes.
▶▶ Use proportional reasoning to solve a problem.
▶▶ Interpret and use ratio in a range of contexts.

Proportional reasoning is useful in many questions.
Always approximate before you calculate.

Use the approximation to check
that your answer is sensible.

example

A garden centre sells a range of conifers.
The table shows last week's sales.

What percentage of conifers sold were Growhii?

Conifer	Sales
Green fir	28
Growhii	156
Pine tree	75
Needelum	12
Grohigherthan	4

You want to find the percentage of conifers sold.
Your answer will be a percentage.

You know that 156 Growhii were sold.
You need to know the total conifers sold.
Approximate: Total conifers sold ≈ 300
                Proportion of Growhii sold ≈ $\frac{150}{300} = \frac{1}{2} = 50\%$
Calculate: Total conifers sold = 28 + 156 + 75 + 12 + 4 = 275
                Proportion of Growhii sold = $\frac{156}{275}$ = 156 ÷ 275 × 100%
                             = 56.7% (1 dp)

A sketch can help you understand the problem.

example

The radius of a circle is 19 mm.
The ratio of the area of the circle to the area of a square is 2 : 1.
What is the length of side of the square (to the nearest mm)?

Area of circle = 2 × area of
square
Area of square = $\frac{1}{2}$ × area of circle

You want to know the length of side of a square.
You know that radius = 19 mm.
Your answer will be in millimetres.

Area of the circle: = πr^2
Approximate: $3 \times 20^2 = 3 \times 400 = 1200$ mm^2
Calculate: $3.14 \times 19^2 = 1133.54$ mm^2

Area = length^2 Area = 2 × area of square

Area of the square: = length^2
Approximate: $\frac{1}{2}$ area of circle ≈ $\frac{1}{2} \times 1200$ mm^2 = 600 mm^2
Calculate: $\frac{1}{2} \times 1133.54$ mm^2 = 566.77 mm^2

Length of the square: = $\sqrt{\text{area}}$
Approximate: $\sqrt{600} \approx 25$ mm
Calculate: $\sqrt{\text{area of square}} = \sqrt{566.77}$ mm^2 = 24 mm (to the nearest mm)

Exercise P1.4

1   Shoaib, Imran and Inzimam share
    £12 600 in the ratio of their ages.
    Shoaib is $1\frac{1}{2}$ times as old as Imran.
    Inzimam is 7 years older than Imran.
    Shoaib is 15 years old.
    Calculate how much money each person
    receives.

2   A farmer decides to divide her field into
    sections for planting different crops.
    She uses $\frac{1}{6}$ of the field for potatoes.
    She uses $\frac{1}{4}$ of the field for carrots.
    She uses $\frac{5}{9}$ of the field for swedes.
    a   In total, what fraction of the field does
        she not plant with crops?
    b   Each year the cost to rent a similar
        sized field is £150 for each $\frac{1}{36}$ of the
        field.
        i   How much would it cost to rent
            $\frac{2}{9}$ of the field?
        ii What fraction of the field could the
            farmer rent for £1350?

3   The table shows the tins of cat food sold
    in one week in Katsave supermarket.

Cat food	Number of tins sold	Takings
Kitty Chunks	145	£68.15
Kat–u–Fat	570	£222.30
Purrfect Pieces	48	£31.20
Meow!	17	£14.28
Katsave own brand	76	£26.60
Total	856	£362.53

    a   What percentage of the total number
        of tins sold was Kat-u-Fat?
    b   What percentage of the total takings
        was Kat-u-Fat?
    c   Kristof says that the percentage of the
        total number of tins sold that are
        Kat-u-Fat should be the same as the
        percentage of the total takings that are
        Kat-u-Fat.
        Explain why this is not true.

4   Kieran collects caps.
    He has red caps, white caps, green caps
    and blue caps.
    K stands for the number of red caps that
    Kieran owns.
    a   Kieran has the same number of white
        caps as red caps.
        How many white caps does Kieran
        own?
    b   Kieran has three times as many green
        caps as white caps.
        How many green caps does Kieran
        own?
    c   Kieran has 7 more blue caps than red
        caps.
        If Kieran has 43 caps, how many does
        he have of each colour?

5   Johann is cooking a vegetable casserole.
    Here is the recipe for 4 people.

    400 g potatoes
    350 g carrots
    2 onions
    180 g mushrooms
    100 ml of vegetable stock

    a   Johann is cooking for 11 people. What
        weight of carrots will he need to use?
    b   He has 500 g of mushrooms in his
        fridge.
        Does he have enough mushrooms to
        make the casserole for 11 people?
        Explain and justify your answer.

6   Judith is driving her car on a motorway.
    The normal speed of the car on the
    motorway is 105 kmph.
    On average, the car uses 5 ml of fuel for
    every 70 metres travelled.
    How many litres of fuel does the car use if
    it travels for one hour at the normal
    speed?

Geometrical reasoning

This spread will show you how to:
- ▶▶ Represent problems and synthesise information.
- ▶▶ Present a concise reasoned argument.
- ▶▶ Identify exceptional cases or counter-arguments.
- ▶▶ Solve problems using properties of angles and polygons.

KEYWORDS

Prism
Net
Angles on a straight line
Angles in a triangle
Order of rotational symmetry
Trial and improvement

To solve a problem efficiently, you need to identify the most appropriate piece of mathematics.

Here are some strategies for solving a geometry problem:

- ▶ Write down the facts you know.
- ▶ Choose an appropriate method of calculation: mental, written or calculator.
- ▶ Give letters to unknown values you need to find.
- ▶ Use trial and improvement.
- ▶ Show your working to help explain your answer.

These strategies are useful for other sorts of problems too.

example

Calculate the missing angles x, y and z in this diagram.

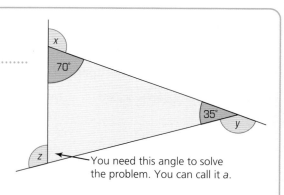

You want to find the **missing angles**.
Your three answers will be in **degrees**.

You know that:

- ▶ sum of angles on a straight line = 180°
- ▶ sum of angles in a triangle = 180°

You need this angle to solve the problem. You can call it a.

Working through the problem mentally:
$x = 180° - 70° = 110°$
      (sum of angles on a straight line = 180°)
$y = 180° - 35° = 145°$
      (sum of angles on a straight line = 180°)
$a = 180° - (70° + 35°) = 75°$
      (sum of angles in a triangle = 180°)
$z = 180° - a = 180° - 75° = 105°$
      (sum of angles on a straight line = 180°)

When calculating with angles, give a reason for each step of your calculation.

Exercise P1.5

1 Look at these shapes:

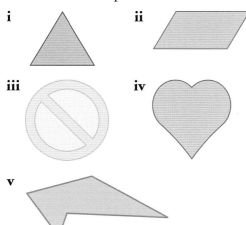

i

ii

iii

iv

v

a Match a correct description to each shape from this list:
  - ▶ 1 line of symmetry; rotational symmetry of order 1
  - ▶ 3 lines of symmetry; rotational symmetry of order 3
  - ▶ 2 lines of symmetry; rotational symmetry of order 0
  - ▶ 0 lines of symmetry; rotational symmetry of order 2
  - ▶ 2 lines of symmetry; rotational symmetry of order 2
  - ▶ 0 lines of symmetry; rotational symmetry of order 1.

b One of the descriptions cannot be matched to any of the shapes above. Explain why it is impossible to draw a shape that fits this description.

2 a Explain exactly how you would map triangle A onto triangle B.

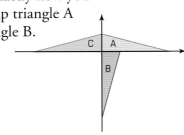

b Explain how you would map triangle B onto triangle C.

3 Here is a triangular prism.

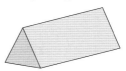

a Which of these nets can be folded to make a triangular prism?

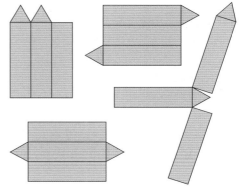

b Explain why one of the nets cannot be folded to make a triangular prism.

4 Wendy is calculating missing angles. Calculate the size of the angles x, y and z.

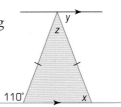

110°

5 Each of these shapes has an area of 20 cm^2.

a Calculate the length of this parallelogram.

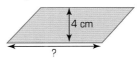

4 cm

?

b Calculate the length and width of this rectangle.

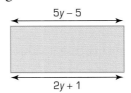

$5y - 5$

$2y + 1$

Checking results

This spread will show you how to:
- ▶▶ Give solutions to an appropriate degree of accuracy.
- ▶▶ Check results using appropriate methods.

KEYWORDS

Mean   Value

Approximation   Expression

Always check your answer. You should:

- ▶ Compare your answer with your approximation.
- ▶ Check that your answer is sensible in the context.
- ▶ Check the accuracy and units required.

Make sure you write your answer in the correct place so that it can be seen clearly.

Here are two other useful checking strategies:

> ▶ You can check an answer by working backwards.

example

The mean of these five cards is 6.
Calculate the value of the missing number.

You want to find the value of a missing card.
Your answer will be a number.

You know:
- ▶ there are 5 cards   and   ▶ the mean of the cards is 6.

The total of the 5 cards = mean × 5 = 6 × 5   = 30
The total of the 4 known cards = 7 + 3 + 9 + 4 = 23
Value of missing card = total of 5 cards – total of 4 known cards
                              = 30 − 23 = 7
Check by working backwards.
Total of 5 cards = 7 + 3 + 9 + 4 + 7 = 30
So the mean of the cards = 30 ÷ 5 = 6

> ▶ You can check an answer using substitution.

example

Simplify $(3d + 5) + (d - 2)$

You want to simplify an algebraic expression.
Your answer will be another algebraic expression.

$(3d + 5) + (d - 2) = 4d + 3$

Check by substituting a value for d in both expressions.
If $d = 1$ $(3d + 5) + (d - 2) = 7$ $4d + 3 = 7$
If $d = 2$ $(3d + 5) + (d - 2) = 11$ $4d + 3 = 11$

The expressions give the same value so the answer is likely to be correct.

Exercise P1.6

1 This is a sketch of an ice-cream wafer in the shape of a sector of a circle.

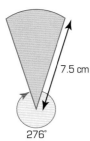

7.5 cm

276°

Make an accurate, full-size scale drawing of this sector.

2 Angie needs to stack 140 tins of beans on the shelves of a supermarket.
A tray of 12 tins of beans weighs 5.28 kg.
What is the weight of the tins Angie has to stack?

3 In a badminton game, each player can score between 0 and 11 points.
Albert plays four games of badminton in a tournament. His mean score for the four games was 9 points.
What points might Albert have scored in each of his four games?

4 Here are two cuboids.

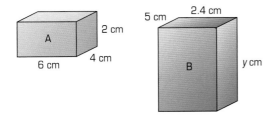

The volume of cuboid A is $\frac{1}{2}$ the volume of cuboid B.
Find the length of the edge marked y.

5 Simplify these algebraic expressions:
a $6 + 4x + 8x$
b $5y + 5 - 3y + 6$

6 Samina and Jack are packing boxes.
They have to pack 1040 tins into the boxes.
Each box can hold 39 tins.
They have already packed 21 boxes.
How many more boxes do they need to pack?

7 ABCD is a rectangle. Calculate the size of the angle p in this rectangle.

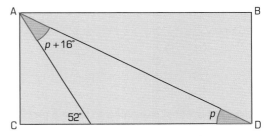

8 This is the graph of $y = 3x - 2$.

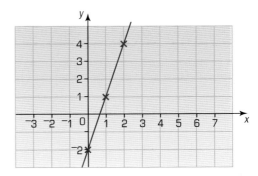

a Does the point (34, 100) lie on the line $y = 3x - 2$?
Explain and justify your answer.
b Write the coordinates of the point which lies on both of the straight lines $y = 2x + 1$ and $y = 6x - 17$.

You should know how to ...

1 Solve substantial problems by breaking them down into simpler steps and tasks.

2 Present a concise, reasoned argument.

3 Give solutions to an appropriate degree of accuracy.

4 Use proportional reasoning to solve a problem.

5 Solve problems using properties of angles and polygons.

Check out

1 A baker is designing a rectangular tray to hold 24 loaves of bread. Each loaf has the dimensions shown.
The tray must not be more than 1 m long. What should the dimensions of the tray be?

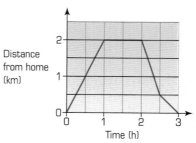 12 cm   24 cm

2 The perimeter of a rectangle is 24 cm. The length of the shortest side is p cm. Show that the area of the rectangle is equal to $12p - p^2$.

3 4 parts of red paint are mixed with 7 parts of yellow paint, to make orange paint.
What is the maximum amount of orange paint that can be made with 100 ml of red paint and 200 ml of yellow paint?

4 A man walks to the park and then returns home. His journey is shown on the graph. Describe his journey.

Distance from home (km)

2

1

0

0   1   2   3
Time (h)

5 Calculate the missing angles in this diagram. Give reasons for your answers.

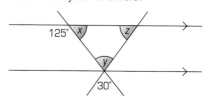
125° x   z

y

30°

Equations, functions and formulae

This unit will show you how to:

▶▶ Distinguish the different roles played by letter symbols in equations, formulae and functions.

▶▶ Simplify or transform algebraic expressions by taking out single-term common factors.

▶▶ Use formulae from mathematics and other subjects.

▶▶ Substitute numbers into equations and formulae.

▶▶ Derive a formula and, in simple cases, change its subject.

▶▶ Construct and solve linear equations.

▶▶ Generate points and plot graphs of linear functions.

▶▶ Solve increasingly demanding problems and evaluate solutions.

▶▶ Represent problems and interpret solutions in algebraic and graphical forms.

▶▶ Move from one form to another to gain a better perspective on the problem.

Medidas circulares • Longitudes y áreas

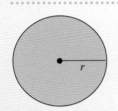

Circunferencia
$L = 2\pi r$

Circulo
$A = \pi r^2$

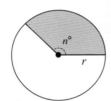

Arco de circunferencia
$L = \dfrac{2\pi r n}{360}$

Sector circular
$A = \dfrac{\pi r^2 n}{360}$

Algebra is the same in many languages.

Before you start

You should know how to ...

1 Substitute values into formulae.

2 Transform simple equations.

$$3x - 4 = 2x + 7$$
$$3x - 2x - 4 = 2x - 2x + 7$$
$$\text{so } x - 4 = 7$$

Check in

1 Copy and complete the table for given values of x when $y = 5 - 3x$.

x	$^{-}2$	$^{-}1$	0	1	2
$y = 5 - 3x$		8			

2 Transform these equations so that the unknown terms are on one side only:

a $3t + 7 = 2t - 5$

b $7m - 12 = 2m + 17$

c $6 - 2y = 3y + 7$

This spread will show you how to:
▶▶ Distinguish the roles played by letters in equations, identities, formulae and functions.
▶▶ Solve linear equations.
▶▶ Substitute numbers into formulae and expressions.

KEYWORDS

Equation	Identity
Formula	Substitute
Function	Variable
Identically equal to ($\equiv$)	

George and Amy are carrying out a Science experiment.

They measure the distance rolled by a metal ball along a glass surface.

Here are their notes:

$$8t + 4 = 28$$
$$4(2t + 1) = 8t + 4$$
$$s = 8t + 4$$
$$t = \text{time}$$
$$s = ut + 4 \qquad s = \text{distance}$$
$$u = \text{speed}$$

▶ The equation $\qquad 8t + 4 = 28 \qquad$ can be solved to find a particular value.
$$8t + 4 - 4 = 28 - 4$$
$$8t = 24$$
$$8t \div 8 = 24 \div 8$$
$$t = 3$$

> This is the balance method – you make sure the sides stay in balance.

▶ The function $\qquad s = 8t + 4 \qquad$ links two variables.

You can substitute the value of one variable to find the value of the other variable:

$$s = 8t + 4$$
$$\text{when } t = 2, \quad s = 8 \times 2 + 4$$
$$s = 18$$

> For every value you substitute there is only one possible solution.

▶ The formula $\qquad s = ut + 4 \qquad$ links two or more variables.
▶ The identity $\quad 4(2t + 1) \equiv 8t + 4 \quad$ is true for every value.
$4(2t + 1)$ is equal to $8t + 4$ no matter what the value of t.

Exercise A5.1

1 Rewrite these expressions in their simplest form:

a $3m + 5m - 6m$ **b** $4m - 5 - 3m + 3$
c $(6 - 2m) + (15 - 3m)$
d $2 \times m \times 3$ **e** $2 \times m \div 3$
f $2(3m + 1) - 8$ **g** $m \times m \div 2$
h $m \div 3$ **i** $m \times 3 - 7$
j $6 \times m \times m \div 3$

2 **a** If $m = 3$ work out the value for each of the expressions in question 1.
b There are eight expressions of equal value when $m = 3$. Write them as equations in pairs, for example, $2m = m + 3$.

3 **a** Solve these two sets of equations:

Set A	**Set B**
i $3a - 4 = 11$	**p** $2p - 4 = p + 8$
ii $16 - 3b = 10$	**q** $2(q + 2) = 11$
iii $15 = 2c + 8$	**r** $3r + 1 = 5r - 9$
iv $d - 7 = 5$	**s** $\frac{s}{3} + 7 = 9$
v $\frac{e}{2} = 3$	**t** $15 - 3t = 4t + 1$

b Match each answer from set A to one from set B.
Write down each pair, for example, $a = s$.

4 Here are six expressions:

a Two are identical for all values of x. Write them out using an identity ($\equiv$) sign.
b When $x = 4$, which other pair of expressions are equal?
c The final two expressions are equal. Write them as an equation and find the value of x.

5 Here are three algebra expressions:

a When $x = 7$, work out the values of each expression.
b If each expression is equal to 13, work out each value for x.
c Two of the expressions are equal. Write all the different equations that could be true and work out the different possible values for x.

6 In a recent Key Stage 3 paper, Paul had his work marked as shown:

> **9** Solve these equations. Show your working.
>
> **i** $4k - 1 = 15$
> $\quad 3k = 15 \qquad k = \underline{5}$ ✗
>
> **ii** $2m + 3 = 10$
> $\quad m + 3 = 5 \qquad m = \underline{2}$ ✗
>
> **iii** $3t + 5 = t + 9$
> $\quad 4t = 14 \qquad t = \underline{3.5}$ ✗
>
> **iv** $2(3n + 2) = 10$
> $\quad 6n + 2 = 10 \qquad n = \underline{2}$ ✗
> $\quad 6n = 12$

a Work out the correct answer for each question.
b Explain what Paul got wrong in each part of the question.

This spread will show you how to:
▶▶ Simplify or transform algebraic expressions by taking out single-term common factors.

KEYWORDS
Expand   Expression
Simplify   Factorise
Multiply out
Take out common factors

Multiplying out brackets is similar to multiplying two numbers:

$$7 \times 67 = 7 \times (60 + 7)$$
$$= 7 \times 60 + 7 \times 7$$
$$= 420 + 49$$
$$= 469$$
$$7(6t + 7) = 7 \times (6t + 7)$$
$$= 7 \times 6t + 7 \times 7$$
$$= 42t + 49$$

×	60	7
7	420	49

×	6t	7
7	42t	49

You can simplify an expression by multiplying out the brackets:

$$3(2t - 1) + 5 - 2t = 3 \times 2t - 3 \times 1 + 5 - 2t$$
$$= 6t - 3 + 5 - 2t$$
$$= 4t + 2$$

This is called **expanding** the brackets.

You can also simplify an expression by taking out a common factor:

The expression $4t + 2$ has two terms: $4t$ and 2.
They have a common factor of 2: $2 \times 2t$ and 2×1
So, taking out the common factor: $4t + 2 = 2(2t + 1)$

This is an identity.

You need to take extra care with negative terms.

example

Simplify $6(3x - 2) - 3(5x - 6)$
Factorise your answer as far as possible.
....
$6(3x - 2) = 6 \times 3x - 6 \times 2$ $^-3(5x - 6) = ^-3 \times 5x - ^-3 \times 6$
$\qquad = 18x - 12$ $\qquad\qquad\qquad = ^-15x + 18$

$6(3x - 2) - 3(5x - 6) = 18x - 12 - 15x + 18$
$\qquad\qquad\qquad = 3x + 6$
$\qquad\qquad\qquad = 3(x + 2)$

You can use factors to help solve problems.

example

If $4x - 10y = 50$, find the value of $2x - 5y$.
....
$2x - 5y = \frac{1}{2}(4x - 10y)$
$\qquad = \frac{1}{2} \times 50 \qquad = 25$

Exercise A5.2

1 In these algebra towers the expression in each cell is made by adding the two expressions below it.

Copy the towers, fill in the missing expressions and write them as simply as possible.

a

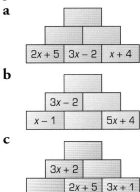

$2x + 5$ | $3x - 2$ | $x + 4$

b

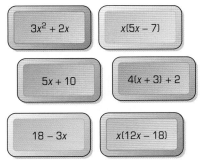

$3x - 2$
$x - 1$ | $5x + 4$

c

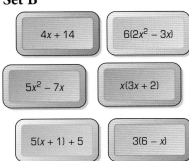

$3x + 2$
$2x + 5$ | $3x + 1$

2 **Equivalent pairs**
Take each expression from Set A and find an equivalent expression in Set B.

Set A

$3x^2 + 2x$ $x(5x - 7)$

$5x + 10$ $4(x + 3) + 2$

$18 - 3x$ $x(12x - 18)$

Set B

$4x + 14$ $6(2x^2 - 3x)$

$5x^2 - 7x$ $x(3x + 2)$

$5(x + 1) + 5$ $3(6 - x)$

3 Multiply out the brackets and simplify these expressions:
a $3(x + 2) + 5$
b $2(3x - 2) + 5$
c $14x - 3(2x + 5)$
d $2(3x + 4) + 3(5x - 2)$
e $5(6 - x) - 2(3x + 4)$
f $x(3x + 2) + 4(3x^2 + 2x)$

4 Factorise these expressions.
The first one is started for you.
a $3x + 6y = 3(\qquad)$
b $2x^2 + 5x =$
c $18x - 12y$
d $5x^2 - 7x =$
e $6x^2 - 8x =$
f $4x^3 - 3x^2 + 9x =$

5 **a** For each of these rectangles, find the area and perimeter:

i

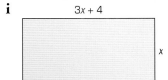

$3x + 4$
x

ii

$6x - 7$
5

b The perimeters of the rectangles are equal.
Find the value of x and use this to find the area of each rectangle.

6 The area of another rectangle is $2x^2 + 6x$.
a Suggest suitable lengths for its sides.
b Find the perimeter of your rectangle.

This spread will show you how to:
▶▶ Construct and solve linear equations.
▶▶ Check results using appropriate methods.

KEYWORDS

Equation   Solution
Solve   Unknown

You can solve even very complicated equations if you take it one step at a time and keep track of your working.

To solve

$$5q - 6 = 3(4q - 9)$$

Multiply out the brackets

$$5q - 6 = 12q - 27$$

Collect the unknowns on one side.
There are more on the right.
Subtract $5q$ from both sides:

$$5q - 5q - 6 = 12q - 5q - 27$$
$$^-6 = 7q - 27$$

Collect the number terms on the other side.
Add 27 to both sides:

$$^-6 + 27 = 7q - 27 + 27$$
$$21 = 7q$$

Divide both sides by 7:
The solution is $q = 3$.

$$21 \div 7 = 7q \div 7$$
$$3 = q$$

Check this works:

left-hand side: $5q - 6 = 5 \times 3 - 6 = 9$
right-hand side: $3(4q - 9) = 3(4 \times 3 - 9) = 3 \times 3 = 9$

They are the same so the solution is correct.

You may need to form an equation before solving it.

example

In this arithmagon, the total in the circle is found by adding adjoining rectangles.

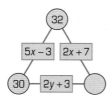

a Find the value of x.
b Find the value of y and the total in the missing circle.

a $5x - 3 + 2x + 7 = 32$
$$7x + 4 = 32$$
$$7x = 28$$
$$x = 4$$

b $5x - 3 + 2y + 3 = 30$
$$20 + 2y = 30$$
$$2y = 10$$
$$y = 5$$

If $x = 4$ and $y = 5$
$2x + 7 + 2y + 3$
$= 8 + 7 + 10 + 3 = 28$

Check this solution by substituting all the values and ensuring the arithmagon works.

Exercise A5.3

1 Solve these equations using a suitable method.
  a $5x + 8 = 23$
  b $3x - 7 = 11$
  c $18 = 4x + 10$
  d $3(2x - 7) = 15$
  e $2(7x - 9) = 10$
  f $5(2x + 3) - 10 = 25$

2 Each of these expressions is equal to $2x + 3$. Form an equation with each expression and solve to find a value for x.
  a $x + 10$ **b** $3x - 8$
  c $5x - 12$ **d** $2(3x - 7)$
  e $5(x + 6)$

3 In an arithmagon, the total in the circle is found by adding adjoining rectangles.
  a

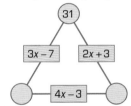

  b

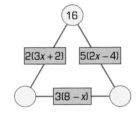

  i For each arithmagon, find the value of x.
  ii Use this value of x to find the totals in the other two circles.

4 Solve these equations using an appropriate method:
  a $5m - 7 = 6$
  b $4(2x + 5) = 34$
  c $9x - 5 = 2(3x + 7)$
  d $3(2 - x) + 4(2x - 3) = 0$
  e $\frac{3x}{4} = 5$ **f** $2 = \frac{11}{x}$

5 Form an algebraic equation for each of these problems and solve it.

  a I multiply a number m by 3 and add 7 to give the same answer as subtracting the number m from 25.

  b I add 6 to a number n and multiply by 7 to give the same answer as subtracting double the number from 60.

  c If I take three times a number y away from 100, I get the same answer as nine times the number add 4.

6 Work out the values of the letters in these equations.

$3a + 5 = 16 - a$

$8 - 3b = 0$

$5c - 3 = 11$

Complete the inequality statement using the values for a, b and c to fill each ☐.

☐ < ☐ < ☐

7 Two of these expressions are equal.

  a Write each of the equations that could be true.
  b Find the different possible values for x.
  c If $^-4 < x < 0$, what is the value of x?

This spread will show you how to:

▶▶ Simplify and transform algebraic expressions using common factors.

▶▶ Add simple algebraic fractions.

▶▶ Solve linear equations.

KEYWORDS
Fraction
Common denominator
Lowest common multiple (LCM)
Equivalent   Cancel

To add or subtract fractions $\frac{1}{3} + \frac{3}{5}$:

▶ Find a common denominator: the lowest common multiple of 3 and 5 is 15.

▶ Find equivalent fractions:

$$\frac{1}{3} \xrightarrow[\times 5]{\times 5} = \frac{5}{15} \qquad \frac{3}{5} \xrightarrow[\times 3]{\times 3} = \frac{9}{15}$$

▶ Add the fractions: $\frac{1}{3} + \frac{3}{5} = \frac{5}{15} + \frac{9}{15} = \frac{14}{15}$

You add and subtract algebraic fractions in the same way.

To add $\frac{x}{6} + \frac{2x}{3}$

▶ The lowest common multiple of 3 and 6 is 6.

▶ Find equivalent fractions: $\frac{2x}{3} = \frac{4x}{6}$

▶ Now add: $\frac{x}{6} + \frac{2x}{3} = \frac{x}{6} + \frac{4x}{6} = \frac{5x}{6}$

You can multiply by the lowest common multiple to get rid of fractions in equations.

> You must multiply every term in the equation.

example

Find the value of x in these equations:

a $\frac{x-4}{3} = \frac{x+2}{4}$

b $\frac{3}{t} + \frac{5}{2t} = 11$

. . . .

a The LCM of 3 and 4 is 12.
Multiply each term by 12 and cancel.

$$\frac{^4\cancel{12}(x-4)}{\cancel{3}_1} = \frac{\cancel{12}^3(x+2)}{\cancel{4}_1}$$

so $4(x-4) = 3(x+2)$

$4x - 16 = 3x + 6$

$4x - 3x - 16 = 3x - 3x + 6$

$x - 16 = 6$

$x - 16 + 16 = 6 + 16$

$x = 22$

b The LCM of t and $2t$ is $2t$.
Multiply each term by $2t$ and cancel.

$$2\cancel{t}^1 \times \frac{3}{\cancel{t}} + {}^1\cancel{2t} \times \frac{5}{\cancel{2t}_1} = 11 \times 2t$$

Simplify:

$6 + 5 = 22t$

$11 = 22t$

$11 \div 22 = 22t \div 22$

$\frac{1}{2} = t$

Remember to check your solutions in each side of the equation:

$\frac{22-4}{3} = \frac{18}{3} = 6$ and $\frac{x+2}{4} = \frac{22+2}{4} = \frac{24}{4} = 6$ $\frac{3}{t} + \frac{5}{2t} = \frac{3}{0.5} + \frac{5}{1} = 6 + 5 = 11$

Exercise A5.4

1 Solve these fraction problems.

a $\dfrac{1}{3} + \dfrac{1}{4}$ **b** $\dfrac{3}{4} - \dfrac{1}{3}$

c $\dfrac{2}{5} + \dfrac{1}{4}$ **d** $\dfrac{3}{10} - \dfrac{1}{5}$

2 Solve these harder fraction problems.

a $\dfrac{a}{3} + \dfrac{a}{4}$ **b** $\dfrac{3b}{4} - \dfrac{b}{3}$

c $\dfrac{2c}{5} + \dfrac{c}{4}$ **d** $\dfrac{3d}{10} + \dfrac{2d}{5}$

3 Match a lowest common multiple with each of these questions and work out the value of x.

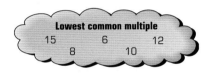

Lowest common multiple

15   6   12   8   10

a $\dfrac{x}{3} - \dfrac{x}{4} = 2$ **b** $\dfrac{x}{5} + \dfrac{x}{2} = 14$

c $\dfrac{2x}{3} - \dfrac{x}{6} = 3$ **d** $\dfrac{x}{3} + \dfrac{2x}{5} = 22$

e $\dfrac{5x}{8} - \dfrac{x}{2} = 5$

4 Find the lowest common multiple to simplify and then solve these equations.

a $\dfrac{x}{2} + \dfrac{x}{3} = 25$ **b** $\dfrac{x+1}{2} + \dfrac{x+2}{3} = 7$

c $\dfrac{x-1}{3} + \dfrac{x+1}{4} = 4$ **d** $\dfrac{x+5}{6} + \dfrac{x-2}{5} = 3$

e $\dfrac{x+5}{2} + \dfrac{x+2}{5} = 5$ **f** $\dfrac{x-2}{3} + \dfrac{x+4}{6} = 7$

5 Solve these equations:

a $\dfrac{x+2}{5} = \dfrac{x-1}{4}$

b $\dfrac{2x+5}{3} = \dfrac{3x+4}{4}$

6 Solve these harder equations:

a $\dfrac{2}{(x+1)} = \dfrac{3}{(x+2)}$

b $\dfrac{4}{(x-1)} = \dfrac{8}{(x+3)}$

> The common denominator of $\frac{1}{(x+1)}$ and $\frac{1}{(x+2)}$ is $(x+1)(x+2)$

7 If x is an integer and $2 \leqslant x \leqslant 5$:

a What different values can x take?

b Substitute integer values for x into this expression:

$\dfrac{16}{x} + \dfrac{12}{2x}$

What are the different values of this expression?

c As the value of x increases, what happens to the value of the expression?

8 If $x = 12$, what is the value of each of these expressions?
Give your answer as a fraction in its simplest form.

a $\dfrac{3}{x} + \dfrac{x-6}{9}$

b $\dfrac{5x+4}{x^2} - \dfrac{16-x}{3x}$

c $\dfrac{4(x-3)}{(x+3)(x-8)}$

Transforming formulae

This spread will show you how to:

▶▶ Use formulae from mathematics and other subjects.

▶▶ Substitute numbers into expressions and formulae.

▶▶ Derive a formula and change its subject.

KEYWORDS

Formula

Rearrange

Subject

The boiling point of water is 100 °C or 212 °F.

The formula $F = \frac{9C}{5} + 32$ changes Celsius to Fahrenheit.

When $C = 100°$, $F = \frac{9 \times \cancel{100}^{20}}{\cancel{5}} + 32$

$= 180 + 32$

$= 212°$

The temperature in parts of England during August 2003 was over 100 °F.

To find 100 °F in Celsius you need to rearrange the formula.

Rearranging formulae is similar to solving equations – you do the same thing to both sides.

example

Rearrange the formula $F = \frac{9C}{5} + 32$ to make C the subject.

Use the formula to change 100 °F to Celsius.

You want the C term on one side. $F = \frac{9C}{5} + 32$

Subtract 32: $F - 32 = \frac{9C}{5} + 32 - 32$

$F - 32 = \frac{9C}{5}$

Multiply by 5: $5(F - 32) = \frac{5 \times 9C}{5}$

$5(F - 32) = 9C$

Divide by 9: $\frac{5(F - 32)}{9} = \frac{9C}{9}$

$\frac{5}{9}(F - 32) = C$

When $F = 100°$, $C = \frac{5}{9}(100 - 32) = 5 \times \frac{68}{9} = 37.8°$ (1 dp)

Exercise A5.5

1 These formulae give the number of lines $T(n)$, in a sequence of patterns.

  i $T(n) = 4n + 4$
  ii $T(n) = 4n + 5$
  iii $T(n) = 3n^2 + 9n$

  a Find $T(n)$ for the third pattern in each sequence ($n = 3$).
  b Find $T(n)$ for each of the formulae when $n = 5$.

2 Here are formulae for the area and volume of different shapes.
Match the correct formula to each shape and find the area or volume of the shapes.

  i $A = \pi r^2$
  ii $A = \frac{1}{2}(a + b)h$
  iii $A = \frac{1}{2}\pi r^2$
  iv $V = \frac{1}{2}\pi r^2 l$
  v $V = \frac{1}{2}bhl$
  vi $V = WLH$

a

b

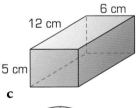

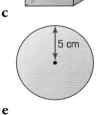

c   **d**

e   **f**

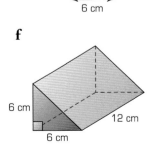

3 Make x the subject of each of these formulae:

  a $a = 2b - x$
  b $a = bx$
  c $a = 5x - 2b$
  d $a = 3bx + 2$
  e $a = \frac{x}{3} - 3b$
  f $a = \frac{x + 2b}{3}$

4 Find the value of x in each equation in question 3 when $a = {}^-2$, $b = 2$.

5 The area of this circle is 154 cm^2.

Use the formula $A = \pi r^2$ to:
  a make r^2 the subject
  b find the value of r^2 (take $\pi = 3.142$)
  c find the value of r.

6 The area of an ellipse is given by $A = \pi ab$.

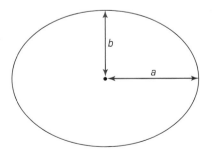

  a Calculate a to 1 decimal place if $A = 30 \text{ cm}^2$ and $b = 3 \text{ cm}$.
  b If $A = 15 \text{ cm}^2$ and $b = 3 \text{ cm}$, work out the length of a.

Solving equations using graphs

KEYWORDS
Graph   Function
Gradient   Intercept
Intersect   Intersection

This spread will show you how to:
▶▶ Generate points and plot graphs of linear functions.
▶▶ Link a graph of an equation to its solution.

You can draw the graph of a function ...   and use it to find pairs of values.

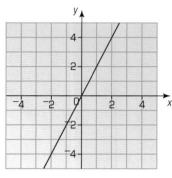

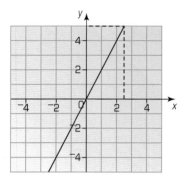

$y = 2x$ When $x = 2.5$, $y = 5$

There are two main methods of drawing the graph of $y = 4 - 3x$.

1 Using a table of values
Plot the pairs and join them.

x	0	1	2	3
$y = 4 - 3x$	4	1	$^-2$	$^-5$

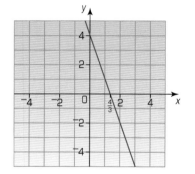

2 Using the gradient and y-intercept

Compare with $y = mx + c$
$y = {}^-3x + 4$
$m = {}^-3$ and $c = 4$

The gradient is $^-3$:
when x goes along 1, y goes down 3.
The y-intercept is 4.

---

example

Find the point at which these two lines intersect: $y = 3x + 1$ and $y = 7 - 3x$

....

The lines intersect at the point that is on both lines –
when the x and y values are the same.

So $3x + 1 = 7 - 3x$
$3x + 3x + 1 = 7 - 3x + 3x$
$6x + 1 = 7$
$6x + 1 - 1 = 7 - 1$ When $x = 1$, $y = 3 \times 1 + 1 = 4$
$6x = 6$ Check in the other equation: when $x = 1$, $y = 7 - 3 \times 1 = 4$
$x = 1$ $(1, 4)$ is on both lines and so is the point of intersection.

Exercise A5.6

1 a Match three of these equations to the lines drawn on the grid.

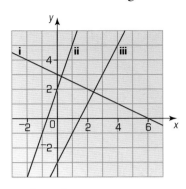

$y = 2x - 3$
$y = \frac{1}{2}x - 3$
$y = 3x + 2$
$y = 3 - \frac{1}{2}x$

b On a similar grid, sketch the line of the fourth equation.

2 Rearrange these linear equations to make y the subject of each equation.

Copy the grid and plot all the equations.

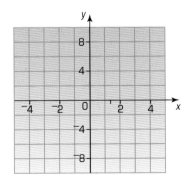

a $y + x - 4 = 0$
b $y - 2x + 3 = 0$
c $\frac{y}{3} - 2 = 0$
d $2y - 3x - 12 = 0$

3 This graph shows the line $y = 5x - 1$.

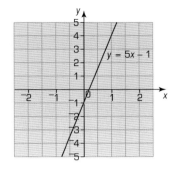

a Does the point $(3, 14)$ lie on the line? Explain how you know.
b Which of these points lie on the line?
$(7, 36)$ $(11, 54)$ $(^-5, ^-24)$ $(^-10, ^-51)$
c Show that the line $y = 2x + 14$ intersects the line $y = 5x - 1$ at the point $(5, 24)$.

4 a Copy the grid and plot at least six points whose x and y coordinates multiply together to make 12 (the point $(6, 2)$ has been plotted for you).

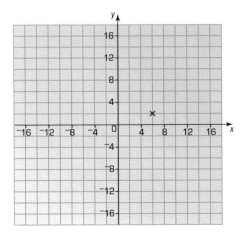

b Carefully draw in the curve $xy = 12$ joining the points you have plotted.
c If $x = ^-1$, what value of y makes $xy = 12$?
d Plot six points where x and y are both negative and $xy = 12$.
e Join up these points to form the second part of the graph for $xy = 12$.

You should know how to ...

1 Construct and solve linear equations using an appropriate method.

2 Use formulae from mathematics and other subjects; substitute numbers into equations and formulae.

3 Generate points and plot graphs of linear functions.

Check out

1 In an arithmagon, the total in each circle is found by adding the rectangles on each side.

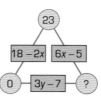

   a Form equations using the arithmagon and find the values of x and y.

   b Work out the missing value in the circle.

2 The volume of this triangular prism is given by:
$$V = \tfrac{1}{2}\, bh\, l$$

   a Rearrange the formula to make the length, l, the subject.

   b This is the front face of a prism.
Its volume is 3600 cm³. Find its length.

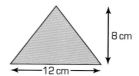

3 **a** Sketch these lines on a grid.
Use the intercept and gradient to help you.

     i $y = 3x - 2$

     ii $y = 8 - 2x$

   b At what point do the two lines intersect?

   c On which line does the point (8, 22) lie?

Statistical reports

This unit will show you how to:

▶▶ Discuss how data relate to a problem and identify possible sources.

▶▶ Construct tables for large sets of raw data.

▶▶ Design and use two-way tables.

▶▶ Gather data from specified secondary sources.

▶▶ Find summary values that represent the raw data, and select the most appropriate statistics.

▶▶ Select, construct and modify suitable graphical representation to progress an enquiry.

▶▶ Interpret graphs and diagrams and draw inferences.

▶▶ Have a basic understanding of correlation.

▶▶ Compare two or more distributions and make inferences.

▶▶ Communicate interpretations and results of a statistical enquiry using selected tables, graphs and diagrams in support.

▶▶ Solve increasingly demanding problems in handling data and evaluate solutions.

▶▶ Present a concise, reasoned argument using symbols, diagrams, graphs and related explanatory text.

▶▶ Represent problems and interpret solutions in graphical form.

▶▶ Identify exceptional cases and counter-examples.

What shall we produce next?

Let's ask our customers.

We'll have to give choices, we can't produce just **anything**.

You have to think carefully before you start to research.

Before you start

You should know how to ...

1 Organise raw data into a frequency table.

2 Calculate the mean, median, mode and range of a set of data.

3 Draw statistical diagrams to represent data.

Check in

1 Some students from class 9B ran a race. Their times were:

17.2 s, 18.3 s, 20.4 s, 16.5 s
19.4 s, 19.2 s, 17.8 s, 21.5 s
18.5 s, 19.2 s, 18.1 s, 17.0 s

Organise this data in a grouped frequency table.

2 Find the mean, median and modal class of the data in question 1.

3 Draw a suitable diagram to represent the data in question 1.

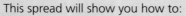

This spread will show you how to:
- ▶▶ Discuss how data relate to the enquiry and identify possible sources.
- ▶▶ Gather data from specified secondary sources.
- ▶▶ Determine the degree of accuracy needed.

KEYWORDS

Survey   Questionnaire
Primary data   Hypothesis
Secondary data
Data collection sheet

Karen is researching cars and pollution.
She collects primary data by carrying out a survey.
She downloads this table of secondary data from the Internet.

Petrol models	cc	CO₂	mpg	Cost	dB	PE
1.3i 3 Door	1299	147	45.6	431	72	N/A
1.4i	1388	158	42.2	465	70	N/A
1.3i	1299	154	43.5	451	72	N/A
1.3i 3 Door	1299	154	43.5	451	72	N/A
1.4i 3 Door	1388	153	44.1	445	70	N/A
1.4i 3 Door	1399	158	42.2	465	70	N/A
1.4i 16V 3 Door	1388	150	44.8	438	71	N/A
1.6i 16V	1596	157	42.8	459	71	N/A
1.6i 16V	1596	164	40.9	480	71	N/A
Diesel models	**cc**	**CO₂**	**mpg**	**Cost**	**dB**	**PE**
1.4 TDCi LX, Zetec, Ghia	1399	120	61.4	333	70	0.034
1.4 TDCi 3 Door LX; Zetec, Ghia	1399	120	61.4	333	70	0.034
1.4 TDCi 3 Door CL	1399	119	62.8	326	70	0.034
1.4 TDCi LX Zetec, Ghia	1399	117	64.2	319	70	0.036
1.4 TDCi 3 Door LX: Zetec, Ghia	1399	117	64.2	319	70	0.034
1.4 TDCi CL	1399	114	65.7	311	70	0.036
1.4 TDCi 3 Door CL	1399	114	65.7	311	70	0.034

Karen may need to round some of the figures to make the data easier to use.

The data is for the Ford Fiesta range. Lots of data are available on the website, but Karen thought that this selection was most useful for her project.

You need to choose data carefully for your project.

- ▶ Primary data is often easy to collect, but is not always representative.
- ▶ Secondary data is often more useful, but you must choose it carefully, and make sure that you know what it all means.

Code	Explanation
cc	engine size
CO₂	carbon dioxide emissions (g/km)
mpg	fuel consumption (miles per gallon)
Cost	fuel costs (£ per 6000 miles)
dB	noise – decibels
PE	particulate emissions (g/km)

Karen decides that the CO_2 figures are most important. She goes on to collect the same information for other cars, and then uses charts to compare the data.

Exercise D3.1

1 Here are some hypotheses that could be tested in a statistical project.

► The level of traffic noise in some classrooms in your school is too high.

► People in the UK own more dogs than people in other parts of Europe.

► School students spend more time watching sport than playing sport.

► Britain's beaches are cleaner now than they were ten years ago.

► Travelling by car is the most common way of getting to school.

For each of these hypotheses:

a Make a list of the things that you would need to know before you could decide whether the hypothesis is correct.

b Explain how you would collect the required data, and whether it would be primary or secondary data, or both. Justify your answers. Include the size and accuracy of your data.

2 Maria wants to find out if there has been a change in the popularity of different holiday destinations.

a She carries out a survey, asking friends where they have gone on holiday. Comment on the likely reliability of Maria's survey results.

b Maria's teacher shows her this data. Explain how Maria could use this data in her project.

Holidays abroad: by destination

United Kingdom			Percentages	
	1971	**1981**	**1991**	**2001**
Spain	34	22	21	28
France	16	27	26	18
Greece	5	7	8	8
United States	1	6	7	6
Italy	9	6	4	4
Irish Republic	–	4	3	4
Other	35	28	31	32
All destinations (=100%) (millions)	4.2	13.1	20.8	38.7

3 Peter decides to test this hypothesis: Fewer children walk to school now than ever before.

► He asks 20 people in his class to fill in a questionnaire.

► They will have to say how they travel to school, and how their parents used to travel to school.

He then collects the data in a data collection sheet.

Do you think that Peter's project will give reliable results? Explain your answer, and suggest any improvements you would make.

4 Decide whether each of these comments is about primary data or secondary data.

a The data is organised in the way you want it.

b This type of data can often be collected quickly and easily.

c The sample size will usually be large enough to be reliable.

d You have to rely on somebody else to collect the data reliably.

e The sample size may be too small to be useful.

f You generally have no control about how the data is organised.

g You should have a good idea of how reliable the data is.

h You may have to spend time finding the data you want, and you might not be able to find it all.

5 Karen's dad says: Diesel cars are noisier than petrol cars, but they go about twice as far on a gallon of fuel. Does the data in the table on page 190 support these views? Explain your answer.

> Sound levels are measured in decibels (dB). The higher the number of decibels, the louder the sound.

KEYWORDS
Discrete   Continuous
Frequency   Raw data
Grouped data
Two-way table
Less than or equal to (≤)

You can organise raw data by sorting it into a table.

Jessica and Jed investigate the traffic flow outside school.
They recorded the number of people in each car and estimated the speed.

This table shows the number of people in the cars.

Number of people	Frequency
1–2	65
3–4	26
5–6	7
7–8	2

This is **grouped discrete** data.
There is no overlap between the groups.

This table shows the speed of the cars.

Speed, v mph	Frequency
$25 \leqslant v < 30$	24
$30 \leqslant v < 35$	42
$35 \leqslant v < 40$	21
$40 \leqslant v < 45$	9

This is **continuous** data. The inequality signs, less than or equal to (≤) and less than (<), ensure that the groups do not overlap.

Using a table makes it easier to interpret the data:

5–6	7

7 cars contained either 5 or 6 people.

$35 \leqslant v < 40$	21

21 cars were travelling between 35 and 40 mph.

You can use a **two-way table** to organise data involving two variables.

Examples of data with two variables include:
▶ age and gender
▶ mode of transport and distance to school.

example

Jessica and Jed thought that there was a difference between the types of vehicle driven by men and women. They collected the data shown in the table.

a What was the most common type of vehicle overall?

b What percentage of the women drivers were driving saloon or estate cars?

Vehicle type	Men	Women
Small hatchback	4	10
Saloon / Estate car	7	12
People carrier / 4x4	4	7
Motorbike / Scooter	2	1
Van / Lorry	3	0

a The most common type of vehicle was 'saloon or estate car', with 19 vehicles.
b 12 out of 30 women drivers were driving saloon or estate cars.
$\frac{12}{30} \times 100\% = 40\%$

Exercise D3.2

1 A technician tests samples of fabric produced by a machine.
She checks 20 samples each day, and records the number of faults in each.
The table shows the numbers of faults recorded over four days.

Grade	Number of faults	Number of samples
Excellent	0–1	37
Satisfactory	2–3	29
	4–5	11
Reject	6–7	3

The numbers of faults in the samples checked the next day were:

0, 4, 1, 2, 1, 0, 1, 0, 6, 2, 2, 3, 0, 0, 1, 1, 5, 1, 1, 0

a Make a new table, and find the total number of samples in each grade for all five days.
b Draw a bar chart to represent the data.
c Draw a pie chart to show the grades awarded to the samples tested over the five days.

2 In an athletics competition, there are four heats in the 100 metre sprint.
The times for the first three heats are shown in the table.

Time (t seconds)	Frequency
$9.9 \leqslant t < 10.0$	1
$10.0 \leqslant t < 10.1$	11
$10.1 \leqslant t < 10.2$	7
$10.2 \leqslant t < 10.3$	3
$10.3 \leqslant t < 10.4$	2

The times (in seconds) for the final heat were:

9.97, 10.03, 10.05, 10.11, 10.15, 10.19, 10.23, 10.31

a Copy the table, and fill in the frequency column for all four heats.
b Draw a frequency diagram to represent the data.

3 Sadie is checking the stock of coloured light bulbs at an electrical store.
There are three colours (red, green and yellow) and three powers (40, 60 and 100 watts).
Sadie writes 40R for a red 40-watt bulb.
Here is her data:

40R	60Y	60G	60G	40G	60G	60Y	60G
60Y	60G	100R	40G	60Y	60R	40R	100Y
60G	40G	60R	60R	60G	60Y	60G	100R
60Y	60R	100G	60G	40R	60Y	100R	60G
60Y	40G	60Y	60R	60Y	60G	60G	40G
60G	40R	60G	60Y	40G	60R	100G	60Y

a Organise the information into a two-way table.
b What percentage of the light bulbs is green?
c If Sadie picks a bulb at random, what is the probability that it will be a green 60-watt bulb?

4 The table shows the number of books taken out of a school library in one term by each of the students in a class.

0	12	4	8	3	9	17	0	4	11
2	7	4	10	11	18	10	8	6	10
4	7	15	14	0	4	5	9	9	8

a Organise the data into a frequency table, using equally sized groups.
b Draw a bar chart to show the data.

5 The table shows the mass (in kilograms to 3 decimal places) of 30 rock samples from a river-bed.

6.575	6.171	8.302	7.753	9.397	7.449	9.720	9.191	6.955	8.604
7.782	9.137	5.319	8.978	8.937	6.431	9.408	6.170	7.882	5.905
7.319	9.740	7.068	10.500	8.118	8.173	7.819	9.335	6.673	8.156

a Organise the data into a frequency table.
b Draw a frequency diagram for the data.

This spread will show you how to:
▶▶ Interpret graphs and diagrams and draw inferences to support or cast doubt on initial conjectures.
▶▶ Have a basic understanding of correlation.

KEYWORDS
Correlation   Relationship
Scatter graph
Variable

When data has two variables, you can use a **scatter graph** to investigate any relationships.

This scatter graph shows the connection between price and engine size for 28 different new cars.

This scatter graph shows the connection between engine size and amount of pollution for the same 28 cars.

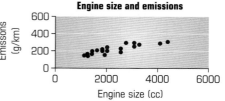

The higher the price, the bigger the engine tends to be.

The larger the engine, the greater the pollution tends to be.

▶ **Correlation** measures how closely connected pairs of variables are.

The graphs both show **positive correlation**.
▶ There is a correlation because the points roughly form a line.
▶ The correlation is positive because as one variable increases, so does the other.

Sets of paired data can also show negative correlation, or no correlation at all.

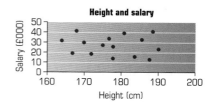

There is a negative correlation between age and price for second-hand cars: as age goes up, price goes down.

There is no correlation between the height of adults and the salary that they earn.

You need more information to decide whether a change in one variable **causes** a change in the other.

This graph shows a positive correlation between the price of a car and the amount of pollution it produces.

However, making a car more expensive does not necessarily cause it to make more pollution!

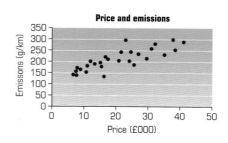

Exercise D3.3

1 Debbie is looking at portable hi-fis in a catalogue.
The table shows details of six different models.

Power (watts)	2	3	5	5	9	10	20
Price (£)	40	50	80	50	110	95	120

  a Plot the data from the table as a scatter graph.

  b Describe the correlation between power and price for these hi-fis.

2 Debbie's dad checks details of vacuum cleaners in the same catalogue.
The table shows his results.

Power (watts)	1300	1400	1400	1500
Price (£)	37	50	60	80

Power (watts)	1600	1600	1300	1800
Price (£)	90	100	50	90

Power (watts)	1100	1200	1400	1600	1500
Price (£)	120	60	90	170	140

  a Plot this data as a scatter graph.

  b Describe any correlation shown by your graph.

3 The table shows the heights of ten boys in Year 9, and the number of children in their families.

Height (cm)	167	172	178	156	168	159	161	162	165	163
Children in family	2	3	3	2	3	2	2	1	2	1

  a Plot a scatter diagram for this data.

  b Is there any correlation between a boy's height and the number of children in his family?
Explain your answer.

4 A group of students did a maths test, a science test, and a sports assessment.
They were given marks out of ten in each one.

Student	A	B	C	D	E	F	G	H	I	J
Maths	4	2	6	8	7	9	10	5	3	8
Science	3	3	7	6	7	8	7	4	5	5
Sports	5	5	2	3	5	8	9	2	8	1

  a Draw a scatter graph to show the relationship between the maths scores and the science scores.

  b Draw another scatter graph to show the relationship between the maths scores and the sports scores.

  c Describe the correlation shown by each of your scatter graphs.

  d Without drawing another graph, predict the correlation between the science scores and the sports scores.

5 Mrs Morgan is Karen's and Ben's maths teacher. She compares their marks (out of 20) for their last five maths homeworks.

Karen's mark	14	13	19	8	17
Ben's mark	14	13	19	8	17

  a Draw a scatter diagram for these marks.

  b What conclusions should Mrs Morgan draw from the graph?
Explain your answer.

6 The table shows the number of sweets and pieces of fruit eaten by eight people in one day.

Fruit	2	1	0	2	5	3	1	2
Sweets	4	8	6	4	2	4	5	6

Draw a scatter graph for this data, and describe any correlation that it shows.

This spread will show you how to:

▶▶ Find summary values that represent the raw data, and select the statistics most appropriate to the problem.

A **skewed distribution** is a set of data with a tendency towards very large or very small values.

Jayne and Nikki recorded the number of people in 10 cars that passed their school at different times one day.

Values that are much larger or smaller than the rest are called **extreme** values.

2 pm	3 pm	4 pm
1, 1, 1, 1, 2, 2, 3, 3, 4, 5	1, 2, 2, 3, 3, 3, 3, 4, 4, 5	1, 2, 3, 3, 4, 4, 5, 5, 5, 5

Jayne and Nikki draw bar charts to show the shape of the distributions:

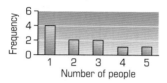

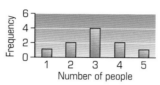

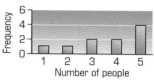

A **skewed** distribution.　　A **symmetric** distribution.　　A **skewed** distribution.

They calculate averages using the raw data:

Mean = 2.3 people	Mean = 3 people	Mean = 3.7 people
Median = 2 people	Median = 3 people	Median = 4 people
Mode = 1 person	Mode = 3 people	Mode = 5 people

▶ The mean and median are the same for the symmetric distribution, but different for the skewed distributions.

example

a   Calculate the mean and median of each of these sets of data.
   A: 5, 5, 5, 5, 5   B: 0, 5, 5, 5, 5   C: 10, 5, 5, 5, 5
b   Comment on the values that you have obtained.

....

a   A: mean = median = 5   B: mean = 4, median = 5   C: mean = 6, median = 5
b   A: The mean equals the median.
    B: One of the 5s reduced to a 0, and the mean has reduced.
    C: One of the 5s increased to a 10 and the mean increased.

The median is unchanged in both skewed distributions.

The mode is the easiest average to calculate, but sometimes it doesn't mean very much!

example

Find the mode of these values: 1, 1, 26, 27, 28, 29 and 30.
Comment on your answer.

....

The mode is 1. This is not a very representative value for the data.

The **modal class** is often a useful average for grouped data.

Exercise D3.4

1 Here are three sets of data:

Set A
17, 20, 21, 18, 16, 20, 21

Set B
15, 14, 78, 16, 15, 13, 15

Set C
48, 52, 46, 3, 3, 45, 47

For each set of data, calculate:
a the mean
b the median
c the mode.
Explain whether each average represents the data fairly.

2 Calculate the mean, median and mode of each of these sets of data.
Explain which average gives the most representative value for each set of data.

Annual salary of 12 people (£1000s)
12, 15, 22, 14, 16, 31, 9, 13, 24, 74, 18, 16

Ages of 20 people in a classroom (years)
14, 13, 13, 13, 14, 14, 13, 14, 52, 14, 13, 14, 28, 14, 14, 14, 13, 14, 14, 13

Number of siblings of 15 people
2, 3, 0, 1, 1, 2, 1, 1, 2, 4, 0, 0, 1, 2, 0

3 Puzzle
Here are five number cards.

a Show how you could choose three of these cards so that the mean of the three numbers is bigger than the median.
b Now pick another three cards to make the median bigger than the mean.
c Finally, choose three cards where the mean and the median are equal.

4 Puzzle
Suki has two number cards. Each card has a whole number written on it.
The mean of the numbers is 8, and the range is 4.
What are the numbers on the cards?

5 Puzzle
Three people have a mean age of 25.
The range of the ages is 8 years, and the median age is 27.
How old is each person?

6 Puzzle
Suneeta has 5 number cards. Each card has a whole number between 1 and 10.
Suneeta says:

> The mode of the numbers is 4. The range is 6 and the mean is 5.

What cards does Suneeta have?
(There are two possible solutions.)

7 The table shows the UK population for different ages groups in 2002.

a Find the modal class for this data.
b Reorganise the data into intervals 0–9 years, 10–19 years, and so on. Find the new modal class.
c Now reorganise the data into intervals covering 20 years each. The first interval should be 0–19 years, and the last one will be '80+'. Find the modal class.

Age	Population (1000s)
0–4	3568
5–9	3833
10–14	3867
15–19	3737
20–24	3652
25–29	4078
30–34	4639
35–39	4775
40–44	4181
45–49	3766
50–54	4042
55–59	3231
60–64	2870
65–69	2575
70–74	2330
75–79	1998
80–84	2381
85–89	757
90+	390

This spread will show you how to:

▶▶ Compare two or more distributions and make inferences, using the shape of the distributions, the range and appropriate statistics.

KEYWORDS

Distribution   Symmetrical
Raw data   Bar-line chart
Range

You can use the shape of graphs to compare distributions.

example

Students from three year groups entered a competition to design a car.
The graphs show the marks awarded.

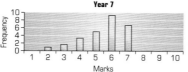

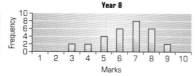

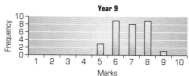

Summarise the main points that the charts show.

| Years 8 and 9 have more students with higher marks – the bars on the right are higher. | The range of marks is greatest for Year 8 (6 marks), and smallest for Year 9 (4 marks). | The shape of the distribution for Year 9 is more symmetrical. |

▶ Year 7 students mostly got marks higher than 5, but there was a long 'tail' of lower marks.
▶ The pattern was similar in Year 8, but the marks were generally higher than in Year 7.
▶ Year 9 students were much more consistent in their scores, and they obtained higher marks than the other two groups.

You can also compare sets of data using the range and an average.

example

The table shows information about marks out of 10 given to two different designs of car.
Use the statistics to compare the marks given to the two designs.

	Mean score	Range
Design A	8.3	9
Design B	5.6	4

Mean score
8.3
5.6

On average, Design A was given higher scores than Design B.

Range
9
4

However, the scores for Design A were more varied – Design B got more consistent scores.

Exercise D3.5

1 A panel of judges awards marks (out of ten) to two different chocolate cakes. The table shows the scores awarded.

Judge	A	B	C	D	E	F	G	H	I	J	K	L
Cake X	9	9	4	9	5	8	4	8	7	3	1	2
Cake Y	6	7	8	5	7	6	7	8	9	7	5	6

For each cake:
a calculate: **i** the mean score
ii the modal score
b draw a bar line chart to show the frequencies.

c Describe and compare the distribution of scores for both cakes.
d Which cake do you think the judges preferred overall? Explain your answer.

2 The head teacher of Grange Park School collects attendance data for three classes in Year 9.
The table shows the number of days with a particular number of absences.
For example, there were five days when there was nobody absent in class 9A.

Absences	0	1	2	3	4	5	6
9A	5	28	24	7	7	3	1
9B	0	5	14	37	16	3	0
9C	2	2	6	7	9	29	20

For each class:
a draw a bar line chart, to show how many days there were with each number of absences
b calculate the mean number of absences
c find the median number of absences
d write down the modal number of absences.

e Describe and compare the distribution of the number of absences for all three classes. You should use your answers to parts **a** to **d** to justify your description.

3 The head office of MegaCorp plc has three computer networks.
The computer manager keeps a record of how many times each network crashes.
The table shows the number of crashes each day for 15 days.

Network A	1	1	2	1	5	0	3	2	1	6	8	7	2	4	3
Network B	3	2	0	0	1	4	2	0	1	1	0	2	2	1	0
Network C	4	3	3	3	4	5	4	3	3	4	3	4	4	3	3

For each network:
a draw a bar line chart to show how often each number of crashes occurred
b work out the mean number of crashes
c find the range of the number of crashes.

d Use your answers to parts **a**, **b** and **c** to compare the reliability of the networks.

4 Two athletics clubs, the Sprinters and the Harriers, enter runners in a marathon.
The times for the runners from each club are shown in the table.

Time, t (hours:minutes)	Sprinters	Harriers
$2{:}30 \leqslant t < 3{:}00$	6	8
$3{:}00 \leqslant t < 3{:}30$	9	15
$3{:}30 \leqslant t < 4{:}00$	10	9
$4{:}00 \leqslant t < 4{:}30$	2	6
$4{:}30 \leqslant t < 5{:}00$	6	4
$5{:}00 \leqslant t < 5{:}30$	0	2
$5{:}30 \leqslant t < 6{:}00$	6	0
$6{:}00 \leqslant t < 6{:}30$	4	0
$6{:}30 \leqslant t < 7{:}00$	2	0

For each athletic club:
a draw a frequency diagram
b find:
i the modal class **ii** the range.

c Compare the results for the two clubs.

This spread will show you how to:
▶▶ Communicate interpretations and results of a statistical enquiry using selected tables, graphs and diagrams in support.

KEYWORDS
Hypothesis   Conclusion
Interpret

The final stage in a statistical project is to explain what your results mean.

Tom is investigating drinking and driving.
His hypotheses are:

▶ the drinking and driving problem is getting worse
▶ drinking and driving mainly affects young male drivers.

Tom interprets his data in a report.

The rate of conviction for drinking and driving accelerated dramatically around 1968, corresponding to the launch of a major government campaign and the introduction of breathalysers.
The rate peaked in the 1980s, and has got slightly better since then.

Drink–driving convictions (thousands)

The table indicates that the vast majority of people convicted of drink–driving offences were male.
However, the percentage of women convicted has increased significantly since 1980.

Year	Female convictions
1960	1.28%
1970	1.31%
1980	3.86%
1990	6.10%
2000	10.40%

The pie chart indicates that drivers in the 17–19 age group are fairly safe. However, there are relatively few of them, and they do not drive as many miles as some other drivers.

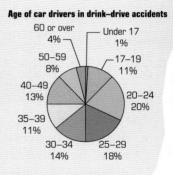

Age of car drivers in drink–drive accidents

- 60 or over 4%
- Under 17 1%
- 50–59 8%
- 17–19 11%
- 40–49 13%
- 20–24 20%
- 35–39 11%
- 25–29 18%
- 30–34 14%

I conclude that young male drivers seem to be a particular problem group for drink–driving.
However, there is little evidence that the problem is getting worse.

There are many other questions that Tom could go on to investigate.

He could try to find out:

▶ why young male drivers are a problem, or
▶ what type of anti-drink-driving campaign is most likely to be effective.

Exercise D3.6

Here are six extracts from statistical reports, each containing a diagram or table and a conclusion.

▶ For each extract, write a paragraph explaining whether you think the conclusion is justified.
▶ If you think that there is a better conclusion, explain carefully what it is.

1 Emergency calls to the ambulance service in England.

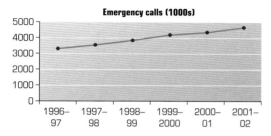

Conclusion: 'People have more accidents every year.'

2 Number of books read by students in class 9J during School Book Month.

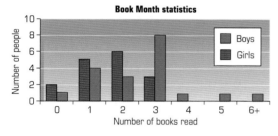

Conclusion: 'Boys are keener on reading books than girls are.'

3 Age of people using the Leisure Centre Swimming Pool (Tuesday 2 pm–4 pm during term-time).

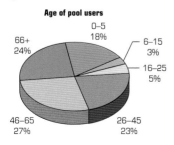

Conclusion: 'Older people take more exercise than younger people.'

4 Finishing times for the school cross-country competition.
Results for Year 8 (Wednesday afternoon) and Year 9 (Thursday morning).

Time, t minutes	Year 8	Year 9
$25 \leqslant t < 30$	4	0
$30 \leqslant t < 35$	26	7
$35 \leqslant t < 40$	35	40
$40 \leqslant t < 45$	11	19
$45 \leqslant t < 50$	4	9
$50 \leqslant t < 55$	0	5

Conclusion: 'Year 8 students are better runners than Year 9 students.'

5 Mean and range of prices of 100 cars sold by two different auction companies.

Company	Mean price	Range
GT Sales	£2850	£4350
A1 Auctions	£2270	£3950

Conclusion: 'GT Sales will get you a better price for your car than A1 Auctions.'

6 Heights and weights of 21 dogs at a show.

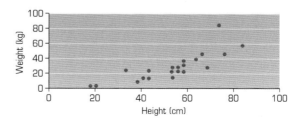

Conclusion: 'The more you feed your dog, the taller it will get.'

You should know how to ...

1 Communicate interpretations and results of a statistical enquiry using selected tables, graphs and diagrams in support.

2 Present a concise, reasoned argument using symbols, diagrams, graphs and related explanatory text.

Check out

1 Here are the test results from a Key Stage 3 Maths test at Brilliant High School:

99, 75, 32, 80, 82, 93, 95,
97, 89, 82, 85, 87, 77, 81.

a Explain which table, graph or diagram you would choose to display the data and why.
b Draw the table or diagram you have chosen.
c Comment on the features of the data.

2 a Explain how the stages of the handling data cycle can be used to help you carry out a statistical project.
Give an example for each stage, using text and diagrams where appropriate.

b The height of an oak tree was measured once a year for 10 years. Here are the results:

0.2 m, 0.5 m, 1.2 m, 1.8 m, 2.2 m,
2.5 m, 2.9 m, 3.1 m, 3.3 m, 3.5 m.

Jon says, 'The tree did not get much larger after seven years.'

Write a paragraph to explain if his conclusion is justified.
Include one diagram or chart.
If you think there is a better conclusion, explain it carefully.

Applying geometrical reasoning

This unit will show you how to:

▶▶ Solve problems using properties of angles, of parallel and intersecting lines, and of triangles and other polygons.

▶▶ Visualise 2-D representations of 3-D objects.

▶▶ Analyse 3-D shapes through 2-D projections.

▶▶ Use and interpret maps and scale drawings.

▶▶ Calculate the surface area and volume of right prisms.

▶▶ Solve increasingly demanding problems and evaluate solutions.

▶▶ Represent problems and synthesise information in geometric form.

▶▶ Present a concise, reasoned argument, using symbols, diagrams and explanatory text.

▶▶ Give solutions to problems to an appropriate degree of accuracy.

You can use properties of shapes to help you design your own kite.

Before you start

You should know how to ...

1 Identify the angle and symmetry properties of triangles and quadrilaterals.

2 Draw nets.

3 Calculate with ratios.

Check in

1 Sketch and describe these shapes:
   a equilateral triangle
   b parallelogram   c kite
   Include angle and symmetry properties.

2 Sketch the net of a cuboid.

3 Reduce these ratios to their simplest terms, without units:
   a 3 kg : 600 g   b 2 hours : 40 minutes
   c 50 cm : 3 km   d 75 cl : 4 l

This spread will show you how to:

▶▶ Solve problems using properties of triangles and other polygons.

▶▶ Understand congruence.

KEYWORDS

Kite   Diagonal

Quadrilateral   Congruent

Line of symmetry

Chloe is designing a kite.

She could make a short kite with an obtuse angle at the base.

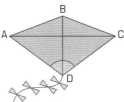

She could make a long kite with an acute angle at the base.

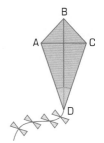

Chloe makes a decision: $\angle A = 120°$ and $\angle D = 40°$.

She calculates the other two angles using her knowledge of quadrilaterals:

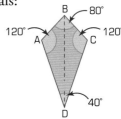

> ▶ A kite has a single line of symmetry.
> So $\angle C = \angle A = 120°$
> ▶ The angles of a quadrilateral add up to $360°$.
> So $\angle B = 360° - (120° + 120° + 40°) = 80°$

Chloe draws in the two rods for her kite.

She calculates the angles that the diagonal makes at A and C:

▶ the diagonals of the kite intersect at right angles

▶ the diagonal BD bisects $\angle B$ and $\angle D$.

$\angle ABM = \angle CBM = 40°$

$\angle ADM = \angle CDM = 20°$

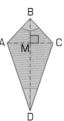

> ▶ You can divide a kite into four right-angled triangles.

So $\angle MAB$
$= 180° - (90° + 40°)$
$= 50°$ (angles of a triangle)

And $\angle MAD$
$= 180° - (90° + 20°)$
$= 70°$ (angles of a triangle)

The two halves of a kite are congruent.

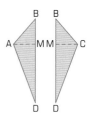

$\angle MCB = \angle MAB = 50°$
$\angle MCD = \angle MAD = 70°$

Exercise S4.1

1 The diagram shows a regular hexagon ABCDEF which is drawn inside a circle with diameter BE.

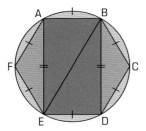

a Find a triangle which is congruent to triangle AEF.
b Find a triangle which is congruent to triangle BED.
c Calculate:
   i ∠AFE   **ii** ∠AEF   **iii** ∠BDE.

2 ABCD is a kite.

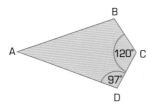

Work out ∠BAD.

3 These triangles are congruent.

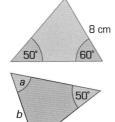

Find the size of:
a angle *a*
b side *b*.

4 Work out all the labelled angles in this kite pattern.
It will help to work them out in order.

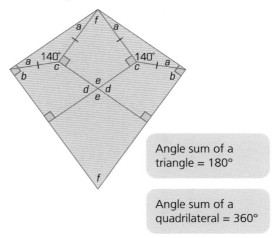

> Angle sum of a triangle = 180°

> Angle sum of a quadrilateral = 360°

5 **Investigation**
   a Explain why a regular pentagon does not tessellate.
   This diagram should help ...

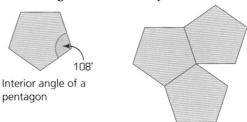

Interior angle of a pentagon

b A square tessellates because the four interior angles marked add to 360°.

Which other regular polygons tessellate? Explain your answers.
   c Do all quadrilaterals tessellate?
   d Do all triangles tessellate?
   Use clear diagrams and angle calculations to explain your answers.
   e Investigate tessellations made using two different regular polygons.

Properties of angles

This spread will show you how to:

▶▶ Solve problems using properties of angles, and of parallel and intersecting lines.

KEYWORDS

Polygon   Prove
Alternate angles
Corresponding angles
Vertically opposite angles

Chloe designs a pattern for her kite.

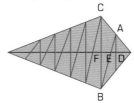

She chooses an angle of 30° at A.

∠B = 30° because it is an **alternate** angle to A.

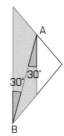

∠C = 30° because it is a **corresponding** angle to A.

In parallel lines:

▶ **Alternate** angles are equal. **(As)**
▶ **Corresponding** angles are equal. **(Cs)**

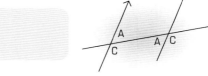

Chloe can calculate ∠AED because she knows that:

▶ Angles in a triangle add up to **180°**.

∠DAE = 30°, and ∠ADE = 90° (right angle)
So ∠AED = 180° − (90° + 30°)
         = 60°

Chloe can calculate ∠FEB because she knows that:

▶ **Vertically opposite** angles are equal.

∠AED = 60°, so ∠FEB = 60°

You can prove that these angle facts are true.

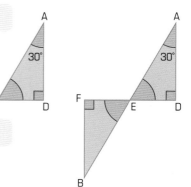

example

Prove that vertically opposite angles are equal.

v and w are vertically opposite.
$v + y = 180°$ (angles on a straight line)
$w + y = 180°$ (angles on a straight line)

So $v = w$

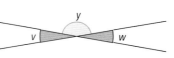

Exercise S4.2

1 Find the angles marked with letters. Give reasons for your answers.

a

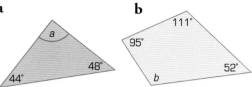

b

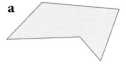

c

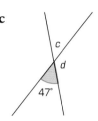

d

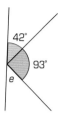

e

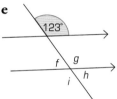

f

g

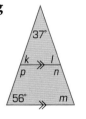

h

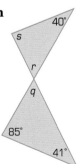

i

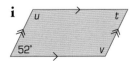

j

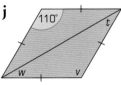

2 Find the **sum** of the interior angles of these polygons.

a   **b**

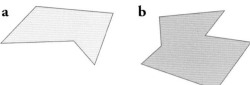

3 Find the interior angle and exterior angle of:
a a regular octagon
b a regular dodecagon (12 sides).

4

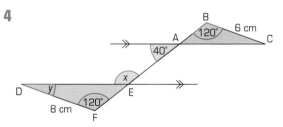

a Calculate angles x and y.
b Show that triangle ABC is similar to triangle DEF.

5 These two polygons are regular.

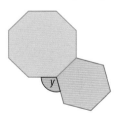

Calculate the angle y.

6 Design a kite with a pattern like the one on page 206.
You will need to work out any angles.

This spread will show you how to:
▶▶ Visualise and use 2-D representations of 3-D objects.
▶▶ Analyse 3-D shapes through 2-D projections, including plans and elevations.

KEYWORDS
Isometric   Cross-section
Plan   Vertical
Elevation   Projection
Net   Solid

Maisie is designing a box kite.

She has drawn the basic shape on **isometric paper**, with the dimensions labelled.

7 cm
10 cm

Always draw vertical lines vertically on isometric paper.

Maisie can also represent her kite using **plans** and **elevations**.

Front elevation

7 cm
10 cm

Side elevation

Plan

▶ Plans and elevations are **projections** of a 3-D solid onto a 2-D surface.

You can identify a 3-D solid from:

▶ its plan and elevations
▶ its net
▶ its cross-sections.

---

example

Here are three views of a 3-D solid.

Front elevation

Side elevation

Plan

a   Identify the solid.
b   Sketch the net of the solid.
c   Sketch two different vertical cross-sections of the solid. What can you say about them?

....

a   The solid is a rectangular-based pyramid.

b
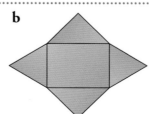

c

The cross-sections are both triangles.

Exercise S4.3

1 Draw these solids on isometric paper:
   a A cube of side 2 cm.
   b A cuboid with sides 4 cm, 3 cm and 2 cm.
   c A hexagonal prism of length 4 cm and cross-section a regular hexagon of side 2 cm.
   d A triangular prism of length 5 cm and cross-section an isosceles triangle of base 2 cm and height 4 cm.
   e A triangular prism of length 6 cm and cross-section an equilateral triangle of side 2 cm.

2 Sketch nets for the 3-D objects in question 1.
   Show all the lengths on your sketches.

3 Use the descriptions in the table to identify the 3-D shapes.

| | Plans and elevations | | |
	Front	Side	Plan
a	circle	circle	circle
b	triangle	triangle	square
c	circle	rectangle	rectangle
d	triangle	triangle	circle

4 You need a box of solids.
   Choose a solid from the box.
   a Name the solid.
   b Draw a front view.
   c Draw a side view.
   d Draw a plan view.
   Repeat for other solids.

   > You can choose a solid from memory if you have no box of solid shapes.

5 These diagrams show the same model from different views.

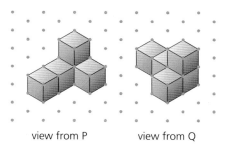

view from P   view from Q

The model is made from 5 cubes.

a One cube is added:

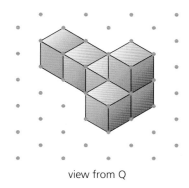

view from Q

   Draw the view from P.

b Now start with the original model (5 cubes).
   Add a cube to a different position.
   i Draw the view from P.
   ii Draw the view from Q.

c Now add two cubes to the original model and draw the views from P and Q.

6 A polyhedron is a 3-D shape with straight edges and flat faces.
   Investigate which polyhedra could have a triangle as a front elevation.

S4.4 Scale drawings and ratio

This spread will show you how to:
- ▶▶ Use and interpret scale drawings.
- ▶▶ Find points that divide a line in a given ratio.

KEYWORDS
Scale drawing
Ratio   Dimensions

Chloe decides how big to make her kite.

She chooses these dimensions:
AD = 70 cm   AB = 30 cm

Because of the symmetry of a kite, it follows that:
CD = 70 cm   CB = 30 cm

Chloe makes a **scale drawing** of her kite.
She chooses a scale of 1 : 10.

Every 1 unit
on the scale
drawing ...

... represents 10
units in reality.

In a scale drawing, the **angles** remain the same.

On the drawing:
AD measures 70 cm ÷ 10 = 7 cm
AB measures 30 cm ÷ 10 = 3 cm

scale 1 : 10

▶ You can express a scale:
- ▶ As a unitary ratio   for example, 1 : 10
- ▶ As a comparison of two quantities   for example, 1 cm to 3 m.

This garden is drawn to a scale of 1 cm to 2 m.

a How long is the actual garden?
b How wide is the actual garden?
c In another rectangular garden, the real length and width are 20 m and 6 m respectively. It is drawn to a scale so that the length on the drawing is 8 cm. Find the width on the drawing.

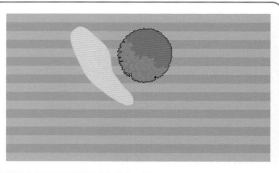

a Length of scaled garden = 7 cm (using a ruler)
7 cm represents 7 × 2 m = 14 m
b Width of scaled garden = 4 cm
4 cm represents 4 × 2 m = 8 m
c 8 cm represents 20 m.
8 cm : 20 m = 8 cm : 2000 cm = 8 : 2000 = 1 : 250
Scaled width = 6 m ÷ 250 = 600 cm ÷ 250 = 2.4 cm

Always:
▶ Convert to the same units.
▶ Cancel a ratio down to its simplest form.

Exercise S4.4

1 **a** On the plan of a house, a door measures 4 cm by 10 cm.
  The plan is drawn to a scale of 1 : 20.
  Find the actual dimensions of the door.

  b Another plan of the same house is made to a
  scale of 1 : 25.
  On the plan, a rectangular window measures 10 cm
  by 5 cm.
  Find the actual dimensions of the window.

2 A dolls' house is an exact replica of a real house,
  on a scale of 1 : 12.
  The dolls' house is 84 cm long, 66 cm wide and 54 cm high.
  Find the dimensions of the real house.

3 A school playground has dimensions of 40 m by 24 m.
  Devlin draws a plan of the playground to a scale of 1 : 200.
  Find the dimensions of the playground in Devlin's plan.

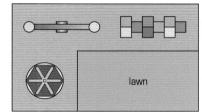

4 The diagram shows an area to be tiled.
  It can be tiled with the blue tiles, or the green tiles.

  a Show that the ratio of blue to green tiles used
  is 1 to 5.

  b Change the measurements of the blue tile so
  that the ratio of blue to green is:

  i 2 : 5

  ii 1 : 8

  iii 1 : 4

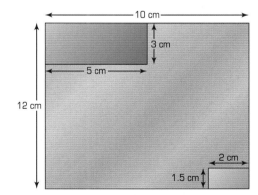

5 The point X divides the line AB in the ratio 1 : 2.
  The ratio has 1 + 2 = 3 parts altogether.
  You divide the horizontal and vertical distance into 3 parts.

  Work out the coordinates of the points which divide:

  a AC in the ratio of 1 : 1

  b AC in the ratio of 1 : 2

  c CB in the ratio of 1 : 2

  d CD in the ratio of 1 : 1

  e AE in the ratio of 1 : 2

> Dividing in the ratio 1 : 1 gives
> the midpoint.

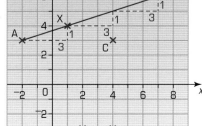

S4.5 Maps and ratio

This spread will show you how to:

▶▶ Use and interpret maps and scale drawings.

▶▶ Interpret and use ratio in a range of contexts.

KEYWORDS

Map   Ratio

Scale

Chloe and Maisie live in the small village of Addersley. The map shows the location of Addersley in relation to the two nearest towns.

Addersley

Brackton

Caleford

The scale of the map is 1 : 80 000.

▶ You can write a map scale as a ratio, without units.

example

Using the map, find the distance:

a from Addersley to Brackton

b from Brackton to Caleford.

Chloe and Maisie want to fly their kites from the top of a hill that is 4.5 km away from Addersley.

c How far from Addersley will the hill be on the map?

....

a Map distance   = 3 cm

  3 cm × 80 000   = 240 000 cm

                      = (240 000 ÷ 100) m

                      = 2400 m

                      = (2400 ÷ 1000) km

  Real distance   = 2.4 km

b Map distance   = 4.3 cm

  4.3 cm × 80 000 = 344 000 cm

                      = (344 000 ÷ 100) m

                      = 3440 m

                      = (3440 m ÷ 1000) km

  Real distance   = 3.44 km

c 4.5 km   = (4.5 × 1000) m

                      = 4500 m

                      = (4500 × 100) cm

                      = 450 000 cm

  450 000 cm ÷ 80 000

                      = 45 cm ÷ 8

                      = 5.625 cm on the map

To convert between metric units of length:

÷ 100   ÷ 1000

cm   m   km

× 100   × 1000

Exercise S4.5

1 This map of Ireland is drawn to a scale of 1 cm to 65 km.

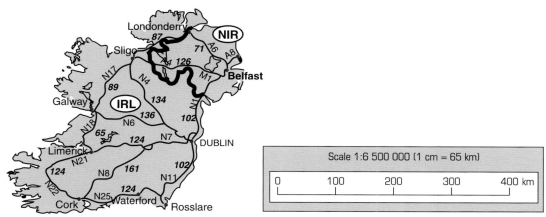

The numbers on the map indicate the distances between major towns.
For example, the distance from Dublin to Belfast is 102 km.
For the following:
i find the distance given by the map.
ii estimate the distance in cm and use the scale to check the numbers given by the map. Are they correct?

a Belfast to Dublin   **b** Limerick to Dublin

c Limerick to Sligo   **d** Londonderry to Cork.

2 The scale of a map is 1 : 20 000.
Find the actual length in metres of 4 cm on the map.

3 The scale of a map is 1 cm to 50 m.
The length of a field is 79.4 m.
What length will the field be on the map?

4 A map is drawn to a scale of 1 : 2000.
Find the actual distance in metres between two farms which are 5.7 cm apart on the map.

5 Copy and complete this table.

	Map scale	Actual length	Map length
a	1 : 20 000	12.5 km	☐ cm
b	1 : 5000	4 km	☐ cm
c	1 : 20 000	☐ km	6.3 cm
d	1 : 5000	☐ km	37 cm
e	1 : 250 000	37 km	☐ cm
f	1 : 250 000	☐ km	12 cm

6 Make an accurate scale drawing of a room in your school, such as your classroom.

S4.6 Surface area and volume of a prism

This spread will show you how to:
▶▶ Calculate the surface area and volume of right prisms.

KEYWORDS
Surface area   Prism
Volume   Accuracy
Net

Joel is making a kite in the form of a hollow hexagonal prism.

Joel wants to estimate how much material he will need.
He needs to calculate the **surface area** of his kite.

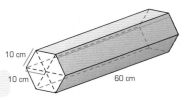

▶ Surface area of a 3-D shape = total area of all its faces

Drawing a net can help you calculate surface area.

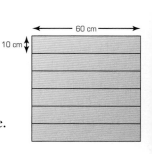

Joel's kite has six rectangular faces (the hexagonal ends are hollow).
Area of each face   = 10 cm × 60 cm
                       = 600 cm^2
Total surface area   = 6 × 600 cm^2 = 3600 cm^2
Joel rounds the answer up to 4000 cm^2 to allow for overlaps and wastage.

Joel wants to calculate the **volume** of his kite.

▶ Volume of a prism = area of uniform cross-section × length

Joel works out the area of each triangle.

He multiplies by the number of triangles.

He multiplies by the length.
This gives the volume of the kite.

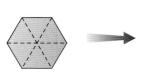

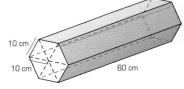

$\frac{1}{2}$ × 10 cm × 8.7 cm
= 43.5 cm^2

43.5 cm^2 × 6 = 261 cm^2

261 cm^2 × 60 cm = 15 660 cm^3

You should give your answer to an appropriate degree of accuracy.

example

The dimensions of a cuboid are 31.3 cm by
20.8 cm by 18.2 cm.
Find the volume of the cuboid, giving your answer
in cm^3 to the nearest whole number.

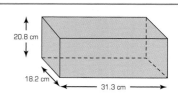

Volume   = 31.3 cm × 20.8 cm × 18.2 cm
          = 11 848.928 cm^3
          = 11 849 cm^3 (to nearest whole number)

Exercise S4.6

1 Find the volume and the surface area of these cuboids.
Use appropriate units for your answers.

a

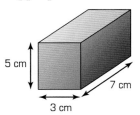

5 cm
7 cm
3 cm

b

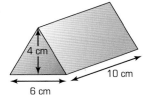

1.3 m
1.7 m
4.9 m

2 Find the volume of these shapes.

a

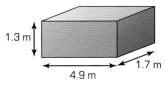

4 cm
10 cm
6 cm

b

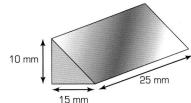

10 mm
25 mm
15 mm

c

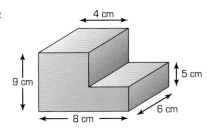

4 cm
9 cm
5 cm
6 cm
8 cm

d

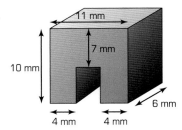

11 mm
7 mm
10 mm
6 mm
4 mm   4 mm

3 Here is a sketch of a swimming pool. Calculate the volume of water the swimming pool holds.

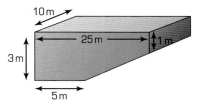

10 m
25 m
1 m
3 m
5 m

4 Find the surface area of this triangular prism.

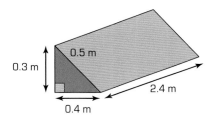

0.5 m
0.3 m
2.4 m
0.4 m

5 This cuboid has a surface area of
(2×8) cm^2 + (2×6) cm^2 + (2×12) cm^2
= $(16 + 12 + 24)$ cm^2 = 52 cm^2

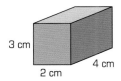

3 cm
4 cm
2 cm

Find another cuboid with the same surface area.

6 **Investigation**
This cube has a volume of 1000 cm^3.

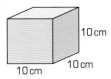

10 cm
10 cm
10 cm

Find other prisms with a volume of 1000 cm^3.

215

You should know how to ...

1 Solve problems using properties of angles, of parallel and intersecting lines, and of triangles and other polygons.

2 Present a concise, reasoned argument, using symbols, diagrams and related explanatory text.

3 Give solutions to problems to an appropriate degree of accuracy.

Check out

1 The diagram shows a circle and a square.

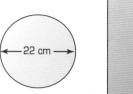 

a The diameter of the circle is 22 cm. What is the area of the circle to the nearest cm²?

b The ratio of the area of the circle to the area of the square is 1 : 3. Find the side length of the square.

2 Here is a plan of a boat crossing a river.

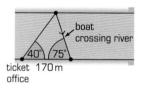

a Use a scale of 1 cm to 20 m to draw an accurate drawing of the boat crossing.

b Work out the actual length of the boat's crossing.
Show your working, and give the units in your answer.

3 Draw the following on a centimetre square grid. Use the line AB as one side of each shape.

A   B

a A right-angled triangle with an area of 8 cm².
b A parallelogram with an area of 8 cm².
c A trapezium with an area of 9 cm².
d An isosceles triangle with an area of 8 cm².

Theoretical and experimental probability

This unit will show you how to:

▶▶ Use the vocabulary of probability in interpreting results involving uncertainty and prediction.

▶▶ Identify all the mutually exclusive outcomes of an experiment.

▶▶ Know that the sum of probabilities of all mutually exclusive outcomes is 1 and use this when solving problems.

▶▶ Estimate probabilities from experimental data.

▶▶ Compare experimental and theoretical probabilities in a range of contexts.

▶▶ Appreciate the difference between mathematical explanation and experimental evidence.

▶▶ Solve increasingly demanding problems in the context of probability and evaluate solutions.

▶▶ Represent problems and synthesise information in graphical form.

You can estimate the probability of a particular outcome when events are random.

Before you start

You should know how to ...

1 Find probabilities based on equally likely outcomes.

2 Calculate with fractions, decimals and percentages.

Check in

1 Find the probability of:
   a choosing a King from an ordinary pack of playing cards.
   b throwing a 5 or a 6 with an ordinary dice.

2 Find:
   a $\frac{2}{3}$ of 45   b 0.3×70 c 15% of 80

This spread will show you how to:
▶▶ Use the vocabulary of probability in interpreting results involving uncertainty and prediction.
▶▶ Identify all the mutually exclusive outcomes of an experiment.

KEYWORDS
Sample space diagram
Outcome   Event
Probability   Independent

A fruit machine has two windows.

Each window can show an apple, a banana, a cherry or a star.

Each picture is equally likely to appear.
You win the top prize if both windows show a star.

The sample space diagram shows all the possible outcomes.

		Window 2			
		Apple	**Banana**	**Cherry**	**Star**
Window 1	**Apple**	AA	AB	AC	AS
	Banana	BA	BB	BC	BS
	Cherry	CA	CB	CC	CS
	Star	SA	SB	SC	SS

▶ There are 16 equally likely outcomes.
▶ There is one outcome that gives the top prize.

The probability of getting the top prize is:

$$p(\text{top prize}) = \frac{\text{number of outcomes resulting in two stars}}{\text{total number of outcomes}}$$

$$= \frac{1}{16}$$

$\frac{1}{16} = 1 \div 16$
$= \textbf{0.0625}$
$= 0.0625 \times 100\%$
$= \textbf{6.25\%}$

▶ You can use a sample space diagram to find probabilities associated with combined events.

You can also work out the probability of getting the top prize by multiplying:

$$p(\text{two stars}) = p(\text{star in window 1}) \times p(\text{star in window 2})$$
$$= \tfrac{1}{4} \times \tfrac{1}{4}$$
$$= \tfrac{1}{16}$$

The result in one window does not affect the result in the other – the two events are independent.

Exercise D4.1

1 Jenny goes to the theatre. She wants to buy a souvenir sweatshirt for her dad as a birthday present.
There are three colours available, and four different pictures.
Jenny cannot decide what to pick, so she decides to choose a colour and a design at random.

Colours	**Pictures**
Black	Large logo
Red	Small logo
Green	Large photo
	Small photo

  a Draw a sample space diagram to show all the combinations of colour and picture that Jenny could pick.

  b Use your diagram to find the probability that Jenny's dad will get a red sweatshirt with a small logo.
  Give your answer as a fraction, a decimal and a percentage.

2 A fair coin is tossed twice.
Show how you can work out the probability of getting two heads.

3 An ordinary dice is rolled twice.
Find the probability of each of these events without using a sample space diagram.

  a The score is 1 both times.

  b Both scores are even numbers.

  c The first score is an odd number, and the second score is a prime number.

4 A bag contains 3 red marbles, 2 green marbles and 5 blue marbles.
Ken picks a marble at random, notes the colour, and puts it back.
He then picks another marble.
Find the probability that:

  a both marbles are red

  b both marbles are blue

  c the first marble is green, and the second one is blue.

5 Explain whether or not the probabilities of success in each pair of trials are independent. The first one is done for you.

	First trial	**Second trial**
a	Tossing a coin to try to get 'heads'.	Tossing a coin to try to get 'tails'.
	These events are independent because the probability of getting heads or tails with the second coin does not depend on the outcome of the first toss.	
b	Rolling a fair dice to try to get a 6.	Rolling the same dice again, trying to get a 4.
c	Trying to burst a balloon with one shot from a peashooter.	Trying to burst the same balloon with another shot from the same peashooter.
d	Trying to pick an even-numbered card at random from a pack marked 1–10.	Trying to pick an even-numbered card from the same pack, after the first card is replaced.
e	Trying to pick an even-numbered card at random from a pack marked 1–10.	Trying to pick an even-numbered card from the same pack, if the first card chosen is not put back in the pack.

Mutually exclusive outcomes

This spread will show you how to:
- ▶▶ Identify all the mutually exclusive outcomes of an experiment.
- ▶▶ Know that the sum of probabilities of all mutually exclusive outcomes is 1 and use this when solving problems.

KEYWORDS
Mutually exclusive
Outcome

▶ Mutually exclusive outcomes cannot occur at the same time.

When you spin a spinner, the scores 3 and 4 are mutually exclusive.

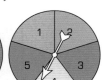

When you toss a coin, heads and tails are mutually exclusive.

At traffic lights, the green man and the red man are mutually exclusive.

example

Amina is about to roll an ordinary fair dice once.

a Write down all the possible outcomes that Amina could get.
b Write down the probability of each outcome.
c Add together all the probabilities. What do you notice?
d What is the probability that Amina will get a 5 or a 6?

....

a The set of mutually exclusive outcomes is: (1, 2, 3, 4, 5 and 6).
b $p(1) = \frac{1}{6}, p(2) = \frac{1}{6}, p(3) = \frac{1}{6}, p(4) = \frac{1}{6}, p(5) = \frac{1}{6}, p(6) = \frac{1}{6}$
c $\frac{1}{6} + \frac{1}{6} + \frac{1}{6} + \frac{1}{6} + \frac{1}{6} + \frac{1}{6} = 1$
The probabilities add up to 1.
d $p(5 \text{ or } 6) = \dfrac{\text{number of favourable outcomes}}{\text{total number of outcomes}}$
$= \frac{2}{6} \quad \text{or} \quad \frac{1}{3}$

The result in part **c** of the example is a general result:

▶ The sum of the probabilities of all mutually exclusive outcomes is 1.

You could have done part **d** of the example in a different way:
$p(5 \text{ or } 6) = p(5) + p(6)$
$= \frac{1}{6} + \frac{1}{6}$
$= \frac{1}{3}$

▶ If two outcomes are mutually exclusive, you can work out the probability of one or the other happening by adding the probabilities.

Exercise D4.2

1 The table shows the probabilities for this five-sided spinner.

Score	1	2	3	4	5
Probability	0.2	0.15	0.15	0.25	0.25

Find the probability that the score is:
a 1 or 2 **b** more than 2
c an odd number.

2 An ordinary dice is rolled.
Explain whether or not these pairs of outcomes are mutually exclusive.
The first one is done for you.

	First outcome	Second outcome
a	The score is 5.	The score is 3.
	These outcomes are mutually exclusive. A dice cannot show a score of 3 and a score of 5 at the same time.	
b	The score is 3.	The score is an even number.
c	The score is an even number.	The score is greater than 4.
d	The score is a prime number.	The score is an even number.
e	The score is a multiple of 5.	The score is a multiple of 3.

3 Every member of a computer club is classified as 'beginner', 'intermediate' or 'expert'.
A member of the club is picked at random.
The table shows the probability of picking each type of member.

Category	Beginner	Intermediate	Expert
Probability	22%		39%

a Copy and complete the table.
b What is the probability of picking a 'beginner' or an 'intermediate' member of the club?

4 A jar contains red marbles, yellow marbles and green marbles. 30% of the marbles are red, and 35% are yellow.
A marble is picked at random.
What is the probability that the marble picked will be yellow or green?
Explain your answer.

5 A set of coloured counters is placed in a bag. The table shows the probability of getting each colour when a counter is picked at random.

Colour	Red	Yellow	Green	Blue	White
Probability	$\frac{1}{3}$	$\frac{1}{6}$	$\frac{1}{4}$	$\frac{1}{6}$	$\frac{1}{12}$

Find the probability of picking:
a a yellow counter or a blue counter
b a yellow counter or a white counter
c a red counter or a green counter
d a yellow counter or a blue counter or a white counter.

6 Martin picks a card at random from this set.

Martin says:

> The probability of picking a yellow card is $\frac{2}{5}$.
> The probability of picking a 3 is $\frac{2}{5}$.
> So the probability of picking a yellow card or a 3 is $\frac{2}{5} + \frac{2}{5} = \frac{4}{5}$.

a Explain why Martin is wrong.
b Work out the correct probability of picking a yellow card or a 3.

7 A raffle ticket is picked at random from a box.
The probability of getting a red ticket is 0.6. The probability of getting a ticket with an even number is 0.5.
Explain why you know that the outcomes 'picking a red ticket' and 'picking an even number' cannot be mutually exclusive.

This spread will show you how to:
▶▶ Identify all the mutually exclusive outcomes of an experiment.
▶▶ Know that the sum of probabilities of all mutually exclusive outcomes is 1 and use this when solving problems.

It is often useful to draw a tree diagram.

Lara has a fair three-sided spinner and an ordinary coin. This tree diagram shows all the possible outcomes when she spins the spinner and tosses the coin.

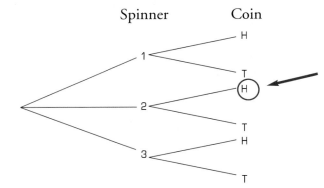

This outcome represents a score of 2 on the spinner, and a head on the coin.

The tree diagram shows that there are six possible outcomes: (1, H), (1, T), (2, H), (2, T), (3, H) and (3, T).

▶ **In a tree diagram, you write outcomes at the ends of branches.**

Lara writes the probabilities of the single events on the tree diagram.

The probability of scoring 1 on the spinner is $\frac{1}{3}$.

The probability of getting a head on the coin is $\frac{1}{2}$.

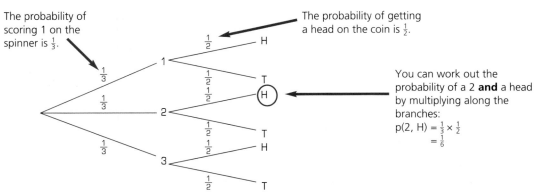

You can work out the probability of a 2 **and** a head by multiplying along the branches:
$p(2, H) = \frac{1}{3} \times \frac{1}{2}$
$= \frac{1}{6}$

▶ **In a tree diagram, you write the probabilities on the branches.**
▶ **You can calculate the probability of a combined event by multiplying along the branches.**

You can use tree diagrams even when the outcomes are not equally likely.

Exercise D4.3

1 Copy and complete the one-stage tree diagrams to show these situations:
  a A fair coin is tossed.

  b An ordinary dice is rolled.

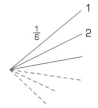

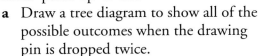
A one-stage tree diagram only has a single set of branches.

2 Draw a two-stage tree diagram to show the possible outcomes when a fair coin is tossed twice.

3 When this drawing pin is dropped on a table, the probability that it lands 'point up' is 0.6.
  a Draw a tree diagram to show all of the possible outcomes when the drawing pin is dropped twice.
  Use your diagram to work out the probability that the drawing pin will land:
  b point up both times
  c point down both times.

4 Marie has two bags, each containing red and blue counters. She picks a counter at random from each bag. The probability of getting a red counter is 0.5 for the first bag, and 0.7 for the second bag.
  a Draw a tree diagram to show all the possible outcomes.
  b Use your diagram to work out the probability that Marie picks two red counters.

5 A four-sided spinner is marked 2, 3, 5 and 7. The probability of each score is shown in the table.

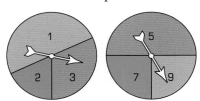

Score, x	2	3	5	7
p(x)	0.4	0.3	0.2	0.1

The spinner is spun twice.
  a Draw a tree diagram to show the possible outcomes.
  The scores for each spin are multiplied together. Use your tree diagram to work out the probability that the product is:
  b less than 5
  c exactly 25
  d more than 40.

6 These two spinners produce the scores shown with the probabilities in the table.

Score on Spinner 1	1	2	3
Probability	0.5	0.15	0.35

Score on Spinner 2	5	7	9
Probability	0.5	0.25	0.25

  a Draw a tree diagram to show the possible outcomes when both spinners are spun.
  Use your diagram to work out the probability that:
  b the product of the scores will be 15
  c the sum of the scores will be 11
  d the difference between the scores will be 8
  e both scores will be even numbers.

This spread will show you how to:

▶▶ Identify all the mutually exclusive outcomes of an experiment.

▶▶ Know that the sum of probabilities of all mutually exclusive outcomes is 1 and use this when solving problems.

▶▶ Estimate probabilities from experimental data.

KEYWORDS

Mutually exclusive
Tree diagram
Experiment
Theoretical probability

You can use a tree diagram to help calculate theoretical probabilities.

---

example

These two spinners have the probabilities shown in the table.

	Red	Blue	Green
Spinner 1	0.25	0.25	0.5
Spinner 2	0.5	0	0.5

Spinner 1   Spinner 2

a Draw a tree diagram.

b Find the probability of getting at least one red when both spinners are spun.

a

```
              0.5   R   p(RR) = 0.125
         R
  0.25       0.5   G   p(RG) = 0.125
         
  0.25       0.5   R   p(BR) = 0.125
         B
             0.5   G   p(BG) = 0.125
  0.5        
             0.5   R   p(GR) = 0.25
         G
             0.5   G   p(GG) = 0.25
```

b The favourable outcomes are:
RR, RG, BR, GR.
They are mutually exclusive, so add.
The probability of getting at least one red is:

p(RR) + p(RG) + p(BR) + p(GR)
= 0.125 + 0.125 + 0.125 + 0.25
= 0.625

If experimental data is available, you can calculate estimates of the experimental probabilities, and compare them with the theoretical values.

example

Ravi used a computer simulation of the experiment in the example above, with 200 trials.

His results are shown in the table.

Result	RR	RG	BR	BG	GR	GG
Frequency	23	25	13	31	63	45
Experimental probability	0.115	0.125	0.065	0.155	0.315	0.225

Estimate the probability of getting at least one red.

Using the experimental data,
p (at least one Red) = 0.115 + 0.125 + 0.065 + 0.315 = 0.620
This agrees very well with the theoretical value of 0.625.

Exercise D4.4

1 Kiran plays two games at a school fair. The probability of winning a prize is 0.3 on the first game, and 0.2 on the second game.
  a Draw a tree diagram to show the possible outcomes.
Use your diagram to find the probability that Kiran gets:
  b two prizes
  c exactly one prize
  d at least one prize.

2 Maya spins a 50p coin, a 20p coin, and a penny.
  a Draw a tree diagram to show the possible outcomes of the experiment.
Use your diagram to find the probability that Maya will get:
  b the same result on each coin
  c at least one tail.

3 These two spinners produce the scores shown with the probabilities in the tables.

Spinner 1

Score	1	2
Probability	0.4	0.6

Spinner 2

Score	3	4
Probability	0.3	0.7

  a Draw a tree diagram to show the possible outcomes when both spinners are spun.
The scores on the two spinners are multiplied together.
Use your diagram to work out the probability that the product will be:
  b exactly 6   **c** 3 or 4
  d an even number   **e** more than 5.

4 Nina has to drive over two level crossings on the way to work each day. The probability that she has to stop is 0.2 at the first one, and 0.3 at the second one.
  a Draw a tree diagram to show the possible outcomes.
Use your tree diagram to work out the probability that the number of times that Nina has to stop at a level crossing will be:
  b zero
  c exactly 1
  d at least 1
  e 2.

5 Jane drops a cup and a saucer on the floor. The probability that the cup breaks is 0.6, and the probability that the saucer breaks is 0.3.
  a Draw a tree diagram to show the possible outcomes.
Use your diagram to work out the probability that:
  b The cup and the saucer both break.
  c Just one of the items breaks.
  d At least one of the items breaks.
  e Neither of the items breaks.

6 Andy flips an ordinary coin three times.
  a Draw a tree diagram to show the possible outcomes.
  b Use your tree diagram to work out the probability of getting each number of 'Heads', from 0 to 3.
  c Draw a bar-line chart to show your results.

7 Carry out the experiment described in question 6.
Estimate the experimental probability for each number of 'Heads'.
Compare your experimental data with the theoretical values you found in question 6.

Comparing theoretical and experimental probability

This spread will show you how to:
- ▶▶ Estimate probabilities from experimental data.
- ▶▶ Compare experimental and theoretical probabilities in a range of contexts.
- ▶▶ Appreciate the difference between mathematical explanation and experimental evidence.

KEYWORDS
Theoretical probability
Predict
Experiment
Expected frequency

You can use theoretical probability to predict the result of an experiment.

The probability of this spinner landing on 'A' is 0.2.
In a game the spinner is spun twice.
You win if you get an 'A' both times.

A tree diagram shows the probability of each outcome.

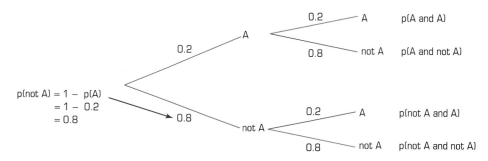

You can work out the probability of winning by multiplying along the branches:
p (A and A) = 0.2 × 0.2 = 0.04

Miriam wants to know how many times she is likely to win in 50 games.
She multiplies the probability by the number of trials (games):
0.04 × 50 = 2
In **theory**, Miriam expects to win twice in 50 games.

▶ Expected frequency = number of trials × probability

Miriam plays the game 50 times and actually wins three times.

▶ The results of an experiment are not always the same as what you expect in theory.

Exercise D4.5

'Less than, more than' is a dice game.
You start by picking one of these number cards.

You then roll an ordinary dice twice.

To win the game:
▶ your first dice score must be **less** than the number on the card
▶ your second dice score must be **more** than the number on the card.

1 Select the card marked '2'.
Copy the tree diagram, and complete it by filling in the
probabilities on each branch.

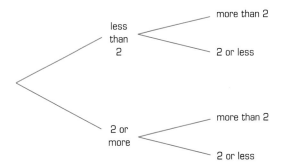

2 Use your tree diagram to work out the theoretical probability
of winning the game when you choose the card marked '2'.

3 Use your answer to question 2 to predict how many times you
would win if you played the game 50 times, choosing the card
marked '2' each time.

4 Now test your prediction by playing the game 50 times, using
an ordinary dice.
You can record your results in a tally chart like this.

Result	Tally	Total
Win		
Lose		

In each game you only need to roll a
second time if the first roll lands on 1.
If it is anything else, you have already lost!

5 Repeat questions 1 to 4 with a different choice of number card.

6 Which card would you choose?
Give a reason for your choice.

Probability experiments

This spread will show you how to:
▶▶ Estimate probabilities from experimental data.
▶▶ Compare experimental and theoretical probabilities in a range of contexts.
▶▶ Appreciate the difference between mathematical explanation and experimental evidence.

KEYWORDS
Theoretical probability
Random   Independent

You can use tree diagrams to work out the theoretical probabilities of combined events, even when the events are not independent.

example

Emily has a set of three cards marked 1, 2 and 3. She picks a card at random, and then picks another card without replacing the first one. Draw a tree diagram to work out the probability of each outcome.

On the first pick, the probability for each card is $\frac{1}{3}$.
The probabilities for the second pick change, because the first card is not replaced: if the first card picked is the 1, then the only cards that can be chosen on the second pick are the 2 and the 3. Each of these has a probability of $\frac{1}{2}$.

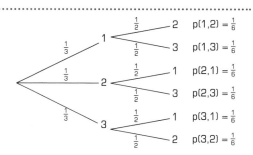

You can work our probabilities for lots of situations in this way.
You must take into account what has already happened at each stage.

example

A bag contains 6 counters: 1 red, 2 green and 3 blue.
A counter is picked at random, and then another one is picked without the first counter being replaced.
Find the probability that at least one blue counter will be picked.

You add together the probabilities for all the outcomes that include at least one blue:

$$\frac{3}{30} + \frac{3}{15} + \frac{1}{10} + \frac{2}{10} + \frac{2}{10}$$

$$= \frac{3}{30} + \frac{6}{30} + \frac{3}{30} + \frac{6}{30} + \frac{6}{30}$$

$$= \frac{24}{30} = \frac{4}{5} = 0.8$$

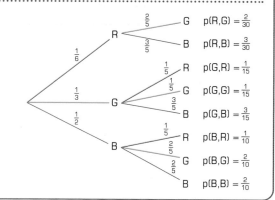

Exercise D4.6

1 The table shows the possible outcomes for the experiment described in the second example on page 228, along with the probabilities for each one.

Outcome	RG	RB	GR	GG	GB	BR	BG	BB
Probability	$\frac{2}{30}$	$\frac{3}{30}$	$\frac{1}{15}$	$\frac{1}{15}$	$\frac{3}{15}$	$\frac{1}{10}$	$\frac{2}{10}$	$\frac{2}{10}$
Probability %	6.7%							

a Copy and complete the table, to show the probabilities as percentages (to one decimal place) in the bottom row.

b Draw a bar-line chart to show the theoretical probability of each outcome.

2 a Carry out an experiment to estimate the experimental probability of each outcome in question 1.
- ▶ Collect counters of the correct colours.
- ▶ Pick one of the counters at random.
- ▶ Pick another counter, without replacing the first one.

Repeat the experiment at least 50 times – the more the better!
Record the results in a tally chart:

Outcome	Tally	Frequency
RG		
RB		
GR		

b Work out the experimental probability of each outcome as a percentage.

c Plot a bar chart to show the experimental probability for each outcome.

> Divide the frequency for the outcome by the total, and convert this to a percentage.

3 Collect together some data for question 2 from several other groups, and combine them into a single large data set.
Use a table like this to collect the data:

Outcome	Frequencies				Total
	Group 1	Group 2	Group 3	Group 4	
RG					
RB					
GR					

> This example uses data from 4 different groups, but you could use more if you have time.

a Work out a new set of experimental probabilities, based on this large data set.

b Draw a bar chart to show your new estimates for the experimental probability of each outcome.

You should know how to ...

1 Know that the sum of probabilities of all mutually exclusive outcomes is 1 and use this when solving problems.

Check out

1 A competition has two rounds. In each round a contestant either wins or loses. The tree diagram shows some of the probabilities.

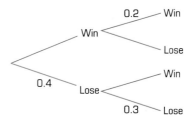

a Copy the tree diagram, and fill in the missing probabilities.

b Use your tree diagram to work out the probability of a contestant winning in **exactly one** of the rounds.

c Find the probability of a contestant winning in both rounds.

2 Compare experimental and theoretical probabilities in a range of contexts.

2 Two four-sided dice have faces labelled 1, 2, 3 and 4. Pran rolls both dice together 100 times and adds the scores.

Here are Pran's results:

Score	2	3	4	5	6	7	8
Probability	6	14	16	24	18	15	8

a Using a sample space diagram, or otherwise, calculate the theoretical probability of each score.

b Use the table to find the experimental probability of each score.

c Compare the theoretical and experimental probabilities and comment on whether the dice are fair.

Trigonometry and Pythagoras' theorem

This unit will show you how to:

▶▶ Use formulae from mathematics.

▶▶ Change the subject of formulae in simple cases.

▶▶ Use a calculator efficiently and appropriately to perform complex calculations.

▶▶ Understand and apply Pythagoras' theorem.

▶▶ Given the coordinates of points A and B, calculate the length of AB.

▶▶ Begin to use sine, cosine and tangent in right-angled triangles to solve problems in two dimensions.

▶▶ Represent problems and synthesise information in algebraic, geometric and graphical form.

▶▶ Give solutions to problems to an appropriate degree of accuracy.

The Ancient Egyptians used properties of triangles to build their pyramids.

Before you start

You should know how to ...

1 Work out a scale factor of enlargement.

2 Calculate squares and square roots.

3 Join coordinate pairs to make a graph.

Check in

1 The diagram shows a shape that has been enlarged. Work out the scale factor of enlargement.

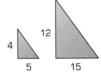

2 Work out:

   a 3^2 **b** 12^2 **c** 6.2^2

   d 17.4^2 **e** $\sqrt{81}$ **f** $\sqrt{196}$

   g $\sqrt{291}$ **h** $\sqrt{46.3}$

3 Plot these points on a grid.

   ($^-2$, $^-16$) ($^-1$, 1) (0, 0) (1, $^-1$) (2, 16)

   Join them to make a smooth curve.

   Describe the curve you have drawn.

KEYWORDS

Scale   Arc
Vertical   Base
Horizontal   Hypotenuse

This spread will show you how to:

▶▶ Begin to use sine, cosine and tangent in right-angled triangles.

The diagram shows a right-angled triangle on a grid.
You can use it to investigate properties of right-angled triangles.

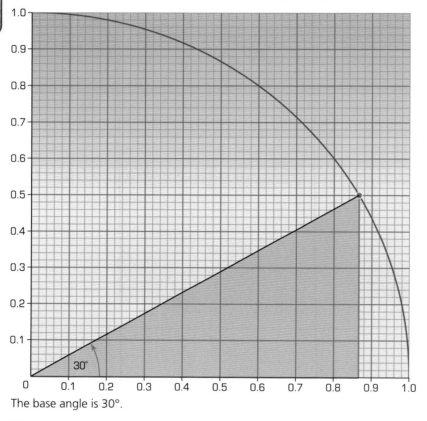

The base angle is 30°.

> The scale used is
> 10 cm = 1 unit.
>
> The arc has radius 10 cm,
> centre (0, 0).
>
> The vertical height is
> marked in green.
>
> The horizontal base is
> marked in red.
>
> The longest side, the
> hypotenuse, is marked in
> black. It measures 1 unit.

You can measure the sides of the triangle using the scale:

Scale length = length in cm ÷ 10

▶ The hypotenuse is always 1 unit long.

For a base angle of 30°:

▶ The vertical height is marked in green.
  The height is 0.5 units.
▶ The horizontal base is marked in red.
  The base is about 0.87 units.

You move the orange dot along the arc to make a different
right-angled triangle with the same hypotenuse.

The hypotenuse is the longest
side in a right-angled triangle.

As the angle increases, the
green line gets longer and the
red line gets shorter.

Exercise B1.1

1 To make an accurate copy of the grid on page 232:

▶ Copy the grid on squared paper, and mark the axes as shown.
▶ Draw a circular arc with a radius of 10 cm, centre (0, 0), using compasses.
▶ Construct a base angle of 30° as shown, using a protractor.
▶ Draw the hypotenuse (black) and vertical (green) height of the triangle.
  The horizontal (red) base of the triangle will lie along the horizontal axis.
▶ Check that the vertical height is 5 cm, and the horizontal base is about 8.7 cm.

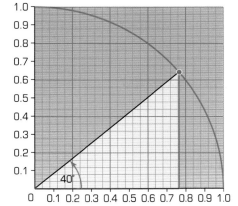

2 a Draw a line with a base angle of 40° on your diagram from question 1.
  Draw the line as far as the arc.
  Draw the vertical side of this triangle.
  b Work out the scale length of the vertical height.
  Scale length = (length in centimetres) ÷ 10
  c Work out the scale length of the horizontal base.

3 Draw more lines on your diagram, for every angle from 0° to 90° in steps of 10°.
For each line, work out (to 2 decimal places) the scale lengths of the horizontal base and vertical height.
Record your results in a copy of this table:

Angle	0°	10°	20°	30°	40°	50°	60°	70°	80°	90°
Horizontal	1.00			0.87						0
Vertical	0			0.50						1.00

4 a Copy this grid.
  ▶ Plot the values for the horizontal base for each angle in question 3:

  (0, 1) (30, 0.87) (90, 0)
  and so on.

  ▶ Join the points with a smooth curve.
  ▶ Comment on your graph.
  b On a new copy of the grid, plot the values for the vertical height.
  Join them with a smooth curve.
  Comment on your graph.

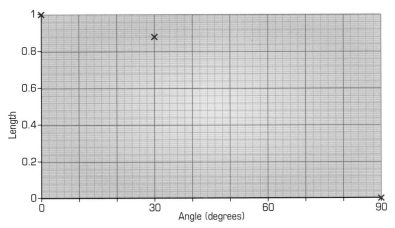

This spread show you how to:

▶▶ Begin to use sine and cosine in right-angled triangles to solve problems in two dimensions.

▶▶ Identify the scale factor of an enlargement.

KEYWORDS

Sine   Enlargement

Cosine   Hypotenuse

Your calculator will tell you the length of the horizontal base and the vertical height when the hypotenuse is 1 unit:

The hypotenuse is the side opposite the right-angle.

example

Find the **vertical height** and the **horizontal base** of this triangle:

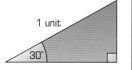

For this triangle:

The vertical height is called the sine of 30°.
sin 30° = 0.5

The horizontal base is called the cosine of 30°.
cos 30° = 0.866 to 3 decimal places

You find the **vertical** height using the 'sin' button. 'sin' is short for **sine.**

You find the **horizontal** base using the 'cos' button. 'cos' stands for **cosine**.

Make sure you can find these values on your calculator.

On some calculators you press:

On other calculators, you press: [sin] [3] [0]

Make sure that your calculator is in **degree** mode.

You can use the idea of enlargements when the hypotenuse is not equal to 1 unit.

This triangle is an enlargement, scale factor 4, of a triangle with a hypotenuse of 1 unit.

For a hypotenuse of 1 unit:

▶ the height is sin 30° = 0.5
▶ the base is cos 30° = 0.866

For a hypotenuse of 4 units:
▶ the height is 4 × sin 30° = 2
▶ the base is 4 × cos 30° = 3.464

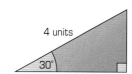

Exercise B1.2

1 a Use a calculator to copy and complete this table. Give lengths to 2 dp.

Angle	Cosine	Sine
0°		
10°		
20°		
30°		
40°		
50°		
60°		
70°		
80°		
90°		

b Compare your results with the table you compiled in question 3 on page 233. What do you notice?

2 Find the lengths of the sides marked with letters.

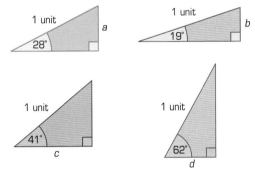

3 Find the lengths of the sides marked with letters.
Remember to include the units.

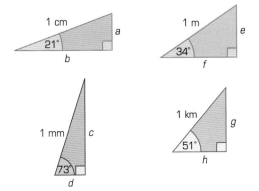

4 Find the lengths marked with letters.

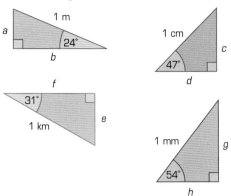

5 Find the lengths marked with letters.

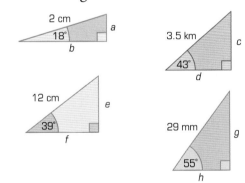

6 Find the lengths marked with letters.

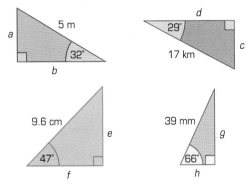

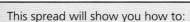

This spread will show you how to:
▶▶ Begin to use sine and cosine in right-angled triangles to solve problems in two dimensions.
▶▶ Use formulae from mathematics.
▶▶ Change the subject in simple cases.

KEYWORDS
Opposite   Inverse
Adjacent   Hypotenuse

In this triangle:
Vertical height = sin 30° × 4
Horizontal base = cos 30° × 4

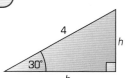

The vertical height is opposite the angle.
When the angle is x, you can say:

▶ Opposite = Sin x × Hypotenuse

or   Sin $x = \dfrac{\text{Opposite}}{\text{Hypotenuse}}$

The horizontal base is adjacent to the angle.
When the angle is x, you can say:

▶ Adjacent = Cos x × Hypotenuse

or   Cos $x = \dfrac{\text{Adjacent}}{\text{Hypotenuse}}$

In these right-angled triangles, an angle is marked with a dot.

Opposite and adjacent are easier terms to use when the triangle is drawn at an angle.

▶ The side **opposite** the angle is green.
▶ The side **adjacent** to the angle is red.
▶ The **hypotenuse** is black.

You can use sines and cosines to find angles given lengths.

example

Find the size of the angle x.

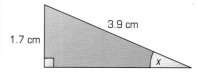

Start with the facts you know:

Opposite = 1.7 cm
Hypotenuse = 3.9 cm

$$\sin x = \frac{\text{opposite}}{\text{hypotenuse}}$$
$$= \frac{1.7 \text{ cm}}{3.9 \text{ cm}}$$
$$= 0.436$$

Use the inverse sine function on your calculator to find the angle:

Press [sin⁻¹] 0.436 = 25.8   (or 0.436 [sin⁻¹] = 25.8)

Angle x = 25.8°

Exercise B1.3

1 Sketch these right-angled triangles.
- ▶ Mark each hypotenuse with an H.
- ▶ Mark the side opposite the angle marked • with an O.
- ▶ Mark the side adjacent to the angle marked • with an A.

The first one is done for you.

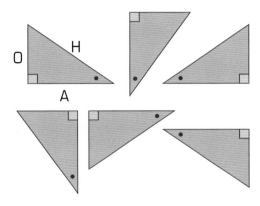

2 Sketch these right-angled triangles. Work out the length of the sides marked with letters.

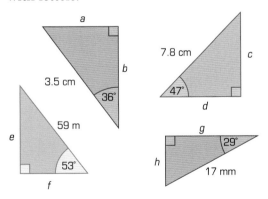

3 Write down a formula you could use to work out the hypotenuse of a right-angled triangle, when you know:
- **a** the size of an angle and the length of the opposite side
- **b** the size of an angle and the length of the adjacent side.

4 Use the formulae you found in question 3 to work out these missing lengths.

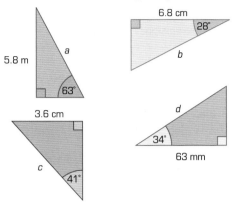

5 Use the inverse sine function ($\sin^{-1}$) on your calculator to find the size of the angles marked with letters. The first one is done for you.

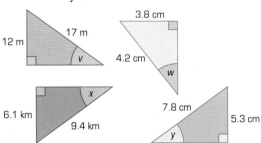

$$\sin v = 12 \div 17 = 0.7059$$
$$\text{so } v = \sin^{-1} 0.7059 = 44.9°$$

6 Use the inverse cosine function ($\cos^{-1}$) on your calculator to find the size of the angles marked with letters in these right-angled triangles.

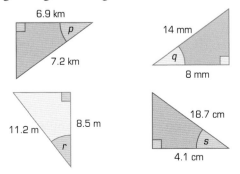

This spread will show you how to:
▶▶ Begin to use tangents in right-angled triangles to solve problems in two dimensions.

KEYWORDS
Sine   Cosine
Hypotenuse   Tangent

For any right-angled triangle, you have a ratio linking:
▶ O and H – the **sine**
▶ A and H – the **cosine**
There is also a ratio linking O and A called the **tangent**.

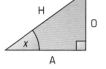

▶ A tangent is a line that just touches a circle.

The grid will help you measure the **tangent**.

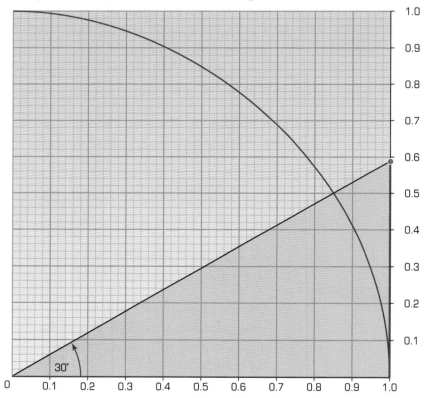

Draw the same grid and arc as on page 232.

Draw the sloping line through the arc to reach the scale on the right-hand side.

The **blue** line shows the length of the **tangent** for an angle of 30°.

tan 30° ≈ 0.58

Whatever angle you draw, the **adjacent** side of the triangle will be 1 unit.

You find the tangent of an angle using this key on a scientific calculator: [tan]

When the adjacent side is 1:
$x = \tan 30°$

When the adjacent side is 5:
$x = 5 \tan 30°$

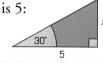

The tangent of an angle can be bigger than 1.

The formula to use when the angle is x is:

▶ Opposite = Tan x × Adjacent   or   Tan $x = \dfrac{\text{Opposite}}{\text{Adjacent}}$

Exercise B1.4

1 Copy and complete the table for the tangents of angles from 0° to 90°. Use a calculator, giving values correct to 3 decimal places.

Angle	Tangent
0°	0
10°	
20°	
30°	
40°	
50°	
60°	
70°	
80°	
90°	–

2 You will not be able to work out the tangent of 90°.
  a Use the diagram on page 238 to explain why it is not possible.
  b Make a table like the one in question 1, to show the tangents of angles from 80° to 89°, in steps of 1°.
  c Use your calculator to work out the tangent of 89.9°.

3 Draw a line graph to show the value of the tangents of angles from 0° to 80° in steps of 10°.

4 Use your calculator to work out the lengths marked with letters.

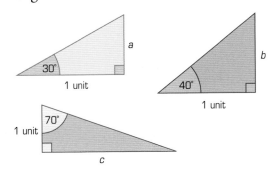

5 Use your calculator to work out the lengths marked with letters.

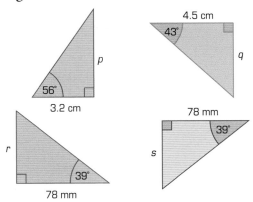

6 Write a formula you could use to work out:
  a the size of an angle, if you know the lengths of the adjacent and opposite sides
  b the length of the adjacent side, if you know the size of an angle and the length of the opposite side.

7 Use the formulae you wrote down in question 6 to find the lengths and angles marked with letters.

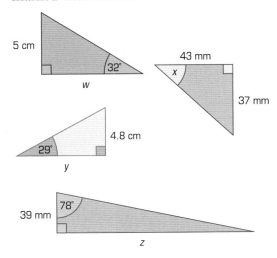

Pythagoras' theorem

This spread will show you how to:

▶▶ Understand and apply Pythagoras' theorem.

▶▶ Given the coordinates of points A and B, calculate the length of AB.

KEYWORDS

Pythagoras' theorem

Opposite   Adjacent

Hypotenuse

You can use sine and cosine to work out the opposite and adjacent sides in a right-angled triangle.

▶ Opposite = $\sin 39° = 0.629$
▶ Adjacent = $\cos 39° = 0.777$

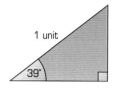

If you square the lengths you get:

▶ Opposite$^2 = 0.629^2 = 0.396$ (to 3 dp)
▶ Adjacent$^2 = 0.777^2 = 0.604$ (to 3 dp)

Opposite^2 + Adjacent^2 = 1

In this triangle,

▶ Opposite = $\sin 47° \times 2 = 1.463$
▶ Adjacent = $\cos 47° \times 2 = 1.364$

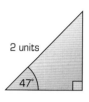

$$\text{Opposite}^2 + \text{Adjacent}^2 = 1.463^2 + 1.364^2$$
$$= 2.140 + 1.860$$
$$= 4$$

Opposite^2 + Adjacent^2 = Hypotenuse^2

> ▶ For any right-angled triangle with sides of length a, b and c (where c is the hypotenuse):
>
> $a^2 + b^2 = c^2$
>
> This is called Pythagoras' theorem.

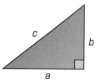

Pythagoras' theorem is extremely useful for finding lengths.

example

Find the lengths of the sides marked by letters.

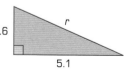

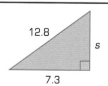

By Pythagoras' theorem:
$5.1^2 + 3.6^2 = r^2$
So $r^2 = 26.01 + 12.96 = 38.97$
So $r = \sqrt{38.97} = 6.24$ to 2 dp

By Pythagoras' theorem:
$s^2 + 7.3^2 = 12.8^2$
So $s^2 = 12.8^2 - 7.3^2$
So $s^2 = 10.51$, and $s = \sqrt{10.51} = 3.24$ to 2 dp

Exercise B1.5

1 To check that a triangle with sides of length 3, 4 and 5 is right-angled:

- ▸ Put $a = 3$, $b = 4$ and $c = 5$ into the equation $a^2 + b^2 = c^2$.
- ▸ Work out the left-hand side of the equation:
  $a^2 + b^2 = 3^2 + 4^2 = 9 + 16 = 25$
- ▸ Work out the right-hand side of the equation: $c^2 = 5^2 = 25$
- ▸ If the two sides of the equation are equal then the triangle must be right-angled.

Follow these steps to check whether these lengths give right-angled triangles:

- **a** $a = 4$, $b = 5$, $c = 6$
- **b** $a = 6$, $b = 8$, $c = 10$
- **c** $a = 5$, $b = 10$, $c = 12$
- **d** $a = 5$, $b = 12$, $c = 13$

2 Find the length of the hypotenuse in each triangle.

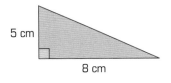

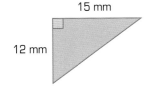

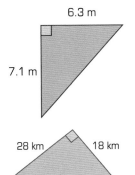

3 Find the length of the sides marked with letters.

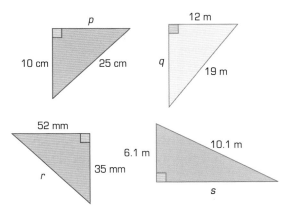

4 To find the distance, d, between two points, draw a right-angled triangle:

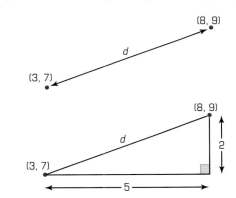

Use Pythagoras' theorem:
$$d^2 = 5^2 + 2^2$$
$$= 25 + 4 = 29$$
so $d = \sqrt{29} = 5.39$

Use Pythagoras' theorem to work out the distance between these points.
It helps to sketch a diagram first.

- **a** $(1, 1)$ and $(4, 5)$
- **b** $(3, 4)$ and $(8, 16)$
- **c** $(2, 8)$ and $(5, 6)$
- **d** $(7, 11)$ and $(12, 5)$
- **e** $(^-1, 3)$ and $(4, 6)$
- **f** $(^-3, ^-2)$ and $(7, 5)$

This spread will show you how to:

▶▶ Understand and apply Pythagoras' theorem.

▶▶ Begin to use sine, cosine and tangent in right-angled triangles to solve problems in two dimensions.

KEYWORDS

Sine   Adjacent
Tangent   Opposite
Cosine   Hypotenuse
Pythagoras' theorem

You can use Pythagoras' theorem and trigonometry to solve problems involving right-angled triangles.

Use Pythagoras' theorem when you know two sides of a triangle, and you want to find the third one.

Use **trigonometry** for problems involving angles.

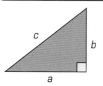

$a^2 + b^2 = c^2$

You need to choose the correct formula to use.
You can use these formula triangles:

Use **sine** if the problem involves the hypotenuse and the opposite side.	Use **cosine** if the problem involves the hypotenuse and the adjacent side.	Use **tangent** if the problem involves the opposite and adjacent sides.

To find opposite, cover O.

$opp = \sin \times hyp$

To find cos, cover C.

$\cos = \dfrac{adj}{hyp}$

To find adjacent, cover A.

$adj = \dfrac{opp}{\tan}$

example

The diagram shows a stretched wire attached to the top of a vertical 10 m pole.
The other end is fixed to the ground, 23 m from the base of the pole.

Find the length, d, of the wire, and the angle x between the wire and the ground.

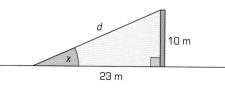

You find the length d by using Pythagoras' theorem.
$$d^2 = 23^2 + 10^2$$
$$= 529 + 100 = 629$$
So $d = \sqrt{629} = 25.08$ m

To find the angle x use tangent, as the adjacent and opposite sides are given in the question.

Using the formula triangle:
$\tan x = 10 \div 23$, so $x = \tan^{-1}(10 \div 23) = 23.5°$

Exercise B1.6

1 Find the angles marked with letters.

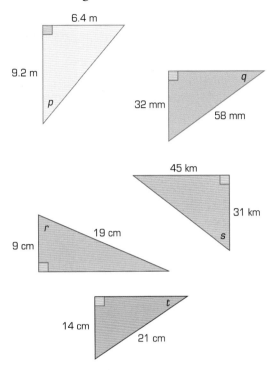

2 Find the lengths of the sides marked with letters.

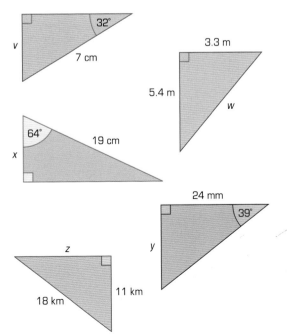

3 A triangle has sides with lengths 6 cm, 8 cm and 10 cm. Without drawing the triangle, prove that it has a right angle.

4 These are the lengths of the sides of some triangles. For each set of lengths, explain whether or not the triangle is right-angled.
   a 7 m, 8 m, 15 m
   b 7 cm, 24 cm, 25 cm
   c 1.5 m, 2 m, 2.5 m

5 A ladder of length 4.5 m leans against a wall of a house so that the top just reaches a windowsill that is 3.7 m above the ground. Work out:
   a The distance, d, between the bottom of the ladder and the base of the wall.
   b The angle, x, between the ladder and the ground.

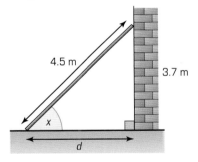

6 A radar station measures the distance to an aeroplane as 9.4 km, and the angle of elevation as 21°. Find the height, h, of the aeroplane above the ground.

You should know how to ...

1 Understand and apply Pythagoras' theorem.

2 Begin to use sine, cosine and tangent in right-angled triangles to solve problems in two-dimensions.

3 Give solutions to an appropriate degree of accuracy.

Check out

1 a A triangle has sides of length 6 cm, 7 cm and 8 cm. Prove that the triangle is not right-angled.

b A right-angled triangle has a hypotenuse of length 2 m. One of the other sides is 1 m long; what is the length of the third side?

c Find the distance between the points whose coordinates are (‾3, 8) and (7, ‾16).

2 Find the marked lengths and angles in these triangles.

a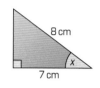

b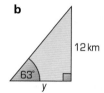

c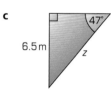

3 Jack worked out the length marked *s* in this triangle.

Jack's answer was *s* = 4.242640687 m.

a Show how Jack found his answer.

b Explain whether Jack's answer is reasonable, and write down the answer that you would give.

accuracy
S4.6

You can round numbers to any degree of accuracy, for example, to the nearest power of 10 or to a given number of decimal places. The degree of accuracy of an answer depends on the information given in the question.

adjacent
B1.3, N2.8, B1.5, B1.6

Adjacent means 'next to'. The side adjacent to the angle in a right-angled triangle is the side next to the angle that is not the hypotenuse.

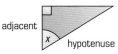

alternate angles
S1.4, S4.2

A pair of alternate angles are formed when a line crosses a pair of parallel lines. Alternate angles are equal.

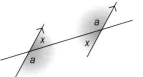

angles in a triangle
P1.5

Angles in a triangle add up to 180°.

angles on a straight line
P1.5

Angles on a straight line add up to 180°.

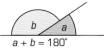

anomaly
D1.4

An anomaly is an exceptional case – a piece of the data which doesn't fit into the general pattern.

approximate, approximation
N2.6, N2.7, P1.4, P1.6

An approximate value is a value that is close to the actual value.

approximately equal to (≈)
N2.6

Approximately equal to means almost the same size.

arc
S1.6, S1.7, B1.1

An arc is a part of a curve.

area: square millimetre, square centimetre, square metre, square kilometre
S2.1, S2.4, S2.5, P1.1

The area of a surface is a measure of the space it covers.

assumed mean
D1.2

An assumed mean is used to simplify the arithmetic when calculating the mean. The assumed mean is subtracted from all the data and added back on once the mean of the smaller numbers has been calculated.

average
D1.2

An average is a representative value of a set of data.

axis, axes
A2.2, D1.6

An axis is one of the lines used to locate a point in a coordinate system.

Glossary

bar chart
D1.3

A bar chart is a diagram that uses rectangles of equal width to display data. The frequency is given by the height of the rectangle.

bar-line chart
D3.5

A bar-line chart is a diagram that uses lines to display data. The lengths of the lines are proportional to the frequencies.

base
A4.3

In index notation, the base is the number which is to be raised to a power, for example, in 5^3, 5 is the base.

base (of plane shape or solid)
B1.1

The lower horizontal edge of a plane shape is called the base. Similarly, the base of a solid is its bottom face.

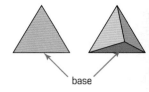
base

best estimate
A3.6

The best estimate of a value is the closest you can achieve.

bias
D2.4

An experiment is biased if not all outcomes are equally likely. A selection is biased if not all the members of the population from which it comes have an equal chance of being chosen

bisect, bisector
S1.9

A bisector is a line that divides an angle or another line in half.

brackets
N1.9, A3.3, N2.8

Operations within brackets should be carried out first.

calculate, calculation
P1.1

Calculate means work out using a mathematical procedure.

cancel, cancellation
N1.1, N2.5, A5.4

A fraction is cancelled down by dividing the numerator and denominator by a common factor,

for example, $\dfrac{24}{40} \overset{\div 8}{\underset{\div 8}{=}} \dfrac{3}{5}$

capacity: litre
S2.5

Capacity is a measure of the amount of liquid a 3-D shape will hold.

centre (of a circle)
S1.6, S2.2

The centre of a circle is the point in the middle.

centre

centre of enlargement
S3.5

The centre of enlargement is the point from which an enlargement is measured.

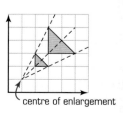
centre of enlargement

centre of rotation S3.1	The centre of rotation is the fixed point about which a rotation takes place.	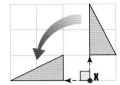
certain D2.1	An event that is certain will definitely happen.	
chord S1.6	A chord is a line joining two points on the circumference of a circle.	
circle S1.8, S2.2, S2.3, S2.4	A circle is a set of points that are the same distance from a fixed point, the centre.	
circumference S1.6, S2.2, S2.3, S2.4, P1.1	The circumference is the distance around the edge of a circle. You calculate it using the formula: $C = 2\pi r$ where r is the radius.	

class interval
D3.4

A class interval is a group that you put data into to make it easier to handle.

common denominator
D2.3, A5.4

A common denominator is an integer that is exactly divisible by all the denominators in a set of fractions, for example, the common denominator of $\frac{2}{3}$ and $\frac{1}{2}$ is 6.

common factor
N2.5

A common factor is a factor of two or more numbers.
For example, 2 is a common factor of 4 and 10.

compare
N1.1, N1.7, D1.5

Compare means to assess the similarity of.

compasses (pair of)
S1.7

Compasses are used for constructions and drawing circles.

compensation
N2.4, N2.5

The method of compensation makes some calculations easier, for example, some multiplications are easier if you double one of the numbers and then compensate by halving the answer.

conclude, conclusion
D3.6

To come to a decision after a series of logical steps.

congruent, congruence
S3.2, S4.1

Congruent shapes are exactly the same shape and size.

construct
A3.2, S1.7

To form an equation from given facts or to draw a line, angle or shape accurately.

construction lines
S1.7

Arcs and lines drawn during the construction of bisectors of angles and lines with a straight edge and compasses.

continuous (data)
D1.2, D3.2

Continuous data can take any value between given limits, for example, less than 1 m.

convention
S1.1, S3.1, S3.2

A convention is an agreed way of describing a situation.
For example, this line is AB:

A ├─────────────────┤ B

Glossary

convert
S2.1

To change.

coordinate pair
A4.5

A coordinate pair is a pair of numbers that give the position of a point on a coordinate grid, for example, (3, 2) means 3 units across and 2 units up.

coordinates
B1.5

Coordinates are the numbers that make up a coordinate pair.

correlation
D3.3

Correlation is a measure of the relationship between pairs of variables.

corresponding angles
S1.4, S3.2, S4.2

A pair of corresponding angles are formed when a straight line crosses a pair of parallel lines. Corresponding angles are equal.

corresponding sides
S3.2, S3.5

Corresponding sides in congruent shapes are equal in length.

cosine
B1.2, B1.4, B1.6

The cosine of an angle in a right-angled triangle is the ratio of the adjacent side a, to the hypotenuse h.

$$\cos x = \frac{a}{h}$$

cost price
N1.5

The price that an item costs to produce.

cross-section
S2.6, S3.4, S4.3

The cross-section of a solid is the 2-D shape you get when you slice the solid perpendicular to its length.

cube, cube number
N1.9, N2.1

A cube number is the product of three equal integers, for example, $27 = 3 \times 3 \times 3$, so 27 is a cube number.

cube root
N1.9, N2.1

The cube root of a number is the value that gives the number when multiplied by itself twice.
For example, $2 \times 2 \times 2 = 8$, so 2 is the cube root of 8, or $2 = \sqrt[3]{8}$.

cubed
A4.3

A number is cubed if it is multiplied by itself twice.
For example, 2 cubed, written 2^3, is $2 \times 2 \times 2 = 8$.

data
D2.4

Data are pieces of information.

data collection sheet
D3.1

A data collection sheet is used to collect data. It is sometimes a list of questions with tick boxes for answers.

decimal place (dp)
N2.3, A3.6

Each column after the decimal point is called a decimal place, for example, 0.65 has two decimal places (2 dp).

decrease
N1.4

To decrease means to make or become smaller.

definition
S1.1

A definition is a set of conditions that uniquely specify something.

degree (°)
S1.1

A degree is a measure of turn. There are 360° in a full turn.

denominator
N1.1

The denominator is the bottom number in a fraction. It shows how many parts there are in total.

derived property
S1.1

A derived property is a consequence of a definition.

diagonal
S4.1

A diagonal of a polygon is a line joining
two vertices but not forming a side.

This is a diagonal.

diameter
S1.6, S2.2, S2.3, P1.1

A diameter is the distance across a circle
through the centre.

diameter

difference
A1.3

The difference between the terms of a linear sequence is constant
(always the same).

digit
N2.2, N2.3, A4.4

A digit is any of the numbers 0, 1, 2, 3, 4, 5, 6, 7, 8, 9.

dimensions
S2.5, S4.4

Flat 2-D shapes have two dimensions:

    length and width   or   base and height

height

base

Solid 3-D shapes have three dimensions:

    length, width and height.

height

width

length

direct proportion
A3.4

Two quantities are in direct proportion if they are always in the
same ratio.

direction
S3.1

The direction is the course, clockwise or anticlockwise, an object
takes when it is rotated.

discrete (data)
D1.2, D3.2

Discrete data can only take certain definite values, for example,
shoe sizes.

distance axis
A2.2

The distance axis is the vertical axis on a distance–time graph.

distance–time graph
A2.2, A2.3, A4.9

A graph showing distance on the vertical axis and time on the
horizontal axis.

distribution
D1.5, D3.4, D3.5

Distribution describes the way data is spread out.

dividend, divisor
N2.7

In this division sum: $6.25 \div 5 = 1.25$

    this is the dividend   this is the divisor   this is the quotient

elevation
S4.3

An elevation is an accurate drawing of the side or front view of a
solid.

enlarge, enlargement
A3.4, S3.5, B1.2

An enlargement is a transformation that multiplies all the sides of a
shape by the same scale factor.

equally likely
D2.1

Events are equally likely if they have the same probability.

equation
A3.1, A3.2, P1.3, A5.1, A5.3

An equation is a statement linking two expressions that are equal in value.

equidistant
S1.6

Equidistant means the same distance from a point or line.

equivalent, equivalence
N1.1, A3.1, N2.5, N2.6, P1.2

Equivalent fractions are fractions with the same value, for example, $\frac{12}{20} = \frac{3}{5}$.

estimate
A3.5, N2.3, N2.6

An estimate is an approximate answer.

evaluate

Evaluate means find the value of an expression.

event
D2.1, D4.1, D4.3

An event is an activity or the result of an activity.

exact, exactly
A3.6

Exact means completely accurate.
For example, three divides into six exactly.

exception
D1.4

An exception is a case which disproves a rule or hypothesis.

expand
A5.2

You expand brackets by multiplying them out, for example, $3(2x - 5) = 6x - 15$.

expected frequency
D2.2, D2.3, D4.5

The expected frequency of an event is the number of times it is expected to occur.

$$\text{Expected frequency} = \text{probability} \times \text{number of trials}$$

experiment
D2.2, D4.4, D4.5

An experiment is a test or investigation to gather evidence for or against a theory.

experimental probability
D2.4

Experimental probability is calculated from the results of an experiment.

explain
P1.2

To explain your answer, you give the reasons why you chose that answer. An explanation may consist of a description of your method or a diagram to illustrate the reason.

expression
A3.1, A3.2, P1.6, A5.2

An expression is a collection of numbers and symbols linked by operations that does not include an equals sign.

exterior angle
S1.2, S1.3

An exterior angle is made by extending one side of a shape.

face
S2.6

A face is a flat surface of a solid.

face

factor
N2.5, A4.1

A factor is a number that divides exactly into another number.
For example, 3 and 7 are factors of 21.

factorise
A5.2

A number or expression is factorised when it is written as a product of its factors, for example, $4a + 6 = 2(2a + 3)$.

fair
D1.6

In a fair experiment there is no bias towards any particular outcome.

favourable outcome
D2.1

A favourable outcome is a successful result of doing something. For example, throwing a 'six' with a fair dice.

formula, formulae
A3.1, S2.4, A5.1, A5.5

A formula is a statement that links variables.

fraction
N1.4, N1.5, N2.4, A5.4

A fraction is a way of describing a part of a whole. For example, $\frac{2}{5}$ of the shape shown is shaded.

frequency
D3.2

Frequency is the number of times something occurs.

frequency diagram
D1.3

A frequency diagram uses bars to display grouped data. The height of each bar gives the frequency of the group, and there is no space between the bars.

frequency table
D3.2

A frequency table shows how often each event or quantity occurs.

function
A2.1, A3.1, A5.1, A5.6

A function is a rule. For example, $+ 2$, $- 3$, $\times 4$ and $\div 5$ are all functions.

function machine
A2.1

A function machine links an input value to an output value by performing the function.

general term
A1.1, A1.2, A1.3, A2.1

The general term of a sequence is an expression which relates its value to its position in the sequence.

generate
A1.1

Generate means produce.

gradient, steepness
A4.5, A4.6, A4.7, A5.6

A gradient is a measure of the steepness of a line.

graph
D1.3, A2.2, A5.6

A graph is a diagram that shows a relationship between variables.

greater than or equal to ($\geqslant$)
N2.3

Greater than or equal to means equal to or more than, for example, $x \geqslant 3$ means x can have any value from 3 upwards.

grouped data
D3.2

When there is a lot of data it is often easier to collect it into groups to see trends more easily. The end of one group must not overlap with the start of the next group:

$$0 < x \leqslant 10 \qquad\qquad 10 < x \leqslant 20$$

This group includes 10.   This group doesn't.

hectare
S2.1

A hectare is a unit of area equal to 10 000 m^2.

highest common factor (HCF)
N1.1, A4.2

The highest common factor is the largest factor that is common to two or more numbers, for example, the HCF of 12 and 8 is 4.

horizontal
B1.1, A4.8, A4.9

Horizontal means level with the flat ground.

Glossary

hypotenuse
S1.7, B1.1, B1.2, B1.3, B1.4, B1.5, B1.6

The hypotenuse is the longest side in a right-angled triangle. It is opposite the right-angle.

hypotenuse

hypothesis, hypotheses
D3.1, D3.6

A hypothesis is an unproved theory.

identity, identically equal to (≡)
A3.1, A5.1

The expressions on either side of an identity are always equal, for example, $3(x + 2) \equiv 3x + 6$ for all values of x.

image
S3.1, S3.2, S3.3

An image is the position of an object following a transformation.

impossible
D2.1

An event is impossible if it cannot happen.

improper fraction
N1.2

In an improper fraction the numerator is greater than the denominator, for example, $\frac{8}{5}$ is an improper fraction.

increase
N1.4, A4.6

To increase means to make or become larger.

independent events
D4.1, D4.6

Two events are independent when one does not affect the outcome of the other.
For example, the outcome from flipping a coin has no effect on the outcome from rolling a dice.
The two events are independent.

index form
N2.1

A number is in index form when it is written as a power: 5^3 is in index form.

index, indices
N1.9, N2.1, N2.2, A4.3

The index of a number tells you how many of the number must be multiplied together. When a number is written in index notation, the index or power is the raised number.
For example, the index of 4^2 is 2. The plural of index is indices.

index laws
A4.3

The rules that tell you how to multiply and divide powers of the same base:
$a^m \times a^n = a^{m+n}$
$a^m \div a^n = a^{m-n}$

index notation
N2.1

A number is written in index notation when it is expressed as a power of another number. For example, 9 in index notation is 3^2.

inference
D1.4

An inference is a conclusion based on evidence.

inspection
A4.7

Inspection is by looking.

integer
N1.2, N1.3, N2.4, A4.4

An integer is a positive or negative whole number (including zero). The integers are: ..., ⁻3, ⁻2, ⁻1, 0, 1, 2, 3, ...

intercept
A4.5, A4.6, A5.6

The intercept is the length between the origin and the point where a line crosses the axis.

interior angle
S1.2, S1.3

An interior angle is inside a shape, between two adjacent sides.

interpret, interpretation
A2.3, D1.1, D1.6, A4.8, D3.6

You interpret data whenever you make sense of it. What you write is your interpretation of the data.

intersect, intersection
A5.6

Two lines intersect at the point where they cross.

intersection

interval
D3.4

An interval is the size of a class or group in a frequency table.

inverse
N1.6, B1.3

An inverse operation has the opposite effect to the original operation, for example, multiplication is the inverse of division.

inverse function
A2.1

An inverse function undoes the effect of a function, for example the function $x \rightarrow 2x - 1$, maps 3 onto 5 and the inverse function $x \rightarrow \dfrac{x + 1}{2}$ maps 5 onto 3.

isometric
S4.3

Isometric grids are designed to make it easier to draw shapes.

justify
P1.2

To justify is to explain or to prove right.

length: millimetre, centimetre, metre, kilometre
S2.1, S2.5

Length is a measure of distance. It is often used to describe one dimension of a shape.

less than or equal to ($\leqslant$)
D3.2

Less than or equal to means equal to or smaller than, for example, $x \leqslant 2$ means that x can have any value from 2 downwards.

line graph
D1.3

On a line graph points are joined by straight lines.

line symmetry
S4.1

A shape has line symmetry if it has a line of symmetry.

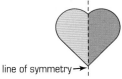

line of symmetry

linear equation, linear expression, linear function, linear relationship
A4.5, A4.7

An equation, expression, function or relationship is linear if the highest power of any variable it contains is 1.
For example, $y = 3x - 4$ is a linear equation, and its graph is a straight line.

linear sequence
A1.3

The terms of a linear sequence increase by the same amount each time.

locus, loci
S1.8, S1.9

A locus is the position of a set of points, usually a line, that satisfies some given condition. Loci is the plural of locus.

lowest common multiple (LCM)
N1.1, A4.2, A5.4

The lowest common multiple is the smallest multiple that is common to two or more numbers, for example, the LCM of 4 and 6 is 12.

map
S3.3, S3.6, S4.5

To map is to follow the rule of a mapping that links two sets of numbers.

Glossary

mapping
A2.1

A mapping is a rule that can be applied to a set of numbers to give another set of numbers.

mean
D1.2, P1.6

The mean is an average value found by adding all the data and dividing by the number of pieces of data.

median
D1.2

The median is an average found by taking the middle value when the data is arranged in size order.

mirror line
S3.1

A mirror line is a line or axis of symmetry.

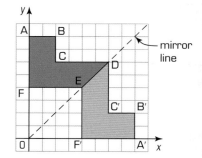

modal class
D1.2, D1.5

The modal class is the most commonly occurring class when the data is grouped. It is the class with the highest frequency.

mode
D1.2

The mode is an average and is the data value that occurs most often.

multiplicative inverse
N1.3, N1.6

You can always get from one number to another by multiplying:
$2 \times 3 = 6$.

$$2 \xrightarrow{\times 3} 6$$
$$\times \tfrac{1}{3}$$

To get back to the original number you can multiply by the multiplicative inverse:

$6 \times \tfrac{1}{3} = 2$.

$\tfrac{1}{3}$ is the multiplicative inverse of 3.

multiply out (expressions)
A5.2

To multiply out a bracket you multiply each term inside by the term outside.
For example, $3(x + 1)$ multiplied out is $3x + 3$.

mutually exclusive
D2.2, D4.2, D4.4

Mutually exclusive events cannot both occur in one trial, for example, if you toss a coin once, you cannot get a Head and a Tail.

nearest
N2.3

Nearest means the closest value.

negative
A3.3

A negative number is a number less than zero.

net
S4.3, S4.6

A net is a 2-D arrangement that can be folded to form a solid shape.

numerator
N1.2

The numerator is the top number in a fraction. It shows how many parts you are dealing with.

object
S3.2

The object is the original shape before a transformation.

operation
A2.1

An operation is a rule for processing numbers or objects. The arithmetic operations are addition, subtraction, multiplication and division.

opposite (sides, angles)
B1.3, B1.5, B1.6

Opposite means across from.

This side is opposite the shaded angle.

order of operations
N1.9, N2.8

The conventional order of operations is:
brackets first, then powers,
then division and multiplication,
then addition and subtraction.

outcome
D2.1, D4.1, D4.2, D4.3

An outcome is the result of a trial.

parallel
S1.4

Two lines that always stay the same distance apart are parallel. Parallel lines never cross or meet.

partition, part
N2.4, N2.5

To partition means to split a number into smaller amounts, or parts. For example, 57 could be split into 50 + 7, or 40 + 17.

percentage (%)
N1.4, N1.5, P1.4

A percentage is a fraction expressed as the number of parts per hundred.

perimeter
S2.2, P1.3

The perimeter of a shape is the distance around it. It is the total length of the edges.

perpendicular
S1.7

Two lines are perpendicular to each other if they meet at a right angle.

perpendicular bisector
S1.8

The perpendicular bisector of a line crosses the line at right angles and cuts it in half.

pi (π)
S2.2, S2.3, S2.4

In a circle, the ratio of the circumference to the diameter is constant. This ratio is called pi, written π.
It doesn't have an exact value – the decimal places go on forever without a pattern.
π is a little over 3.
Good approximations are $\pi = 3.14$ or $\pi = \frac{22}{7}$.

circumference

diameter

pie chart
D1.3

A pie chart is a circle used to display data. The angle at the centre of a sector is proportional to the frequency.

plan
S4.3

A plan of a solid is the view from directly overhead.

Glossary

plane of symmetry
S3.4

A plane of symmetry is a cross-section through a 3-D shape that divides the shape into two identical halves. All prisms have at least one plane of symmetry.

p(*n*)
D2.1

p(*n*) stands for the probability of event *n* occurring.

polygon: pentagon, hexagon, octagon
S1.2, S4.2

A polygon is a closed shape with three or more straight edges.

A pentagon has five sides.

A hexagon has six sides.

An octagon has eight sides.

population pyramid
D1.5

A population pyramid is a back-to-back bar chart comparing two populations.

position-to-term rule
A1.1, A1.3

The position-to-term rule links the value of a term to its position in the sequence.

positive
A3.3

A positive number is greater than zero.

power, index, indices
N1.9, N2.1, N2.2, A4.3, A4.4

When a number is written in index notation, the power or index is the raised number, for example, the power of 3^2 is 2.

predict, prediction
D4.5

Predict means forecast.

primary (data)
D1.1, D3.1

Data you collect yourself is primary data.

prime number
N1.1, A4.1

A prime number is a number that has exactly two different factors.

prime factor
A4.2

A prime factor is a factor that is prime.

prime factor decomposition
A4.1

Expressing a number as the product of its prime factors is prime factor decomposition.
For example, $12 = 2 \times 2 \times 3 = 2^2 \times 3$.

prism
S2.6, S3.4, S4.6

A prism is a 3-D shape with a constant cross-section.

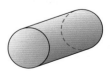

You name a prism by its cross-section.

probability
P1.2, D4.1

Probability is a measure of how likely an event is.

probability scale
D2.1

A probability scale is a line numbered 0 to 1 or 0% to 100% on which you place an event based on its probability.

product
N2.8

The product is the result of a multiplication.

projection
S4.3

When you look at a 3-D shape from different angles you can see 2-D shapes, called projections.
The projections of this shape are:

Front   Side   Plan

proof, prove
S4.2

A proof is a chain of reasoning that establishes the truth of a proposition.

proportion
N1.5, N1.8, S3.5, P1.4

Proportion compares the size of a part to the size of a whole. You can express a proportion as a fraction, decimal or percentage.

proportional to
N1.8, S3.6

Quantities are proportional to one another when they increase or decrease in the same proportion.

pyramid
S3.4

A pyramid is a 3-D shape that tapers to a point called the apex.

You name a pyramid by its base.

Pythagoras' theorem
B1.5, B1.6

In any right-angled triangle, Pythagoras' theorem gives the relationship between the lengths of the sides:

$$a^2 + b^2 = c^2$$

where c is the hypotenuse.

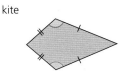

quadrilateral: kite, parallelogram, rectangle, rhombus, square, trapezium
S1.2, S1.5, S2.1, S4.1

A quadrilateral is a polygon with four sides.

rectangle   parallelogram   kite

 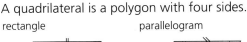

All angles are right angles. Opposite sides equal.

Two pairs of parallel sides.

Two pairs of adjacent sides equal. One line of symmetry.

rhombus   square   trapezium

All sides the same length. Opposite angles equal.

All sides and angles equal.

One pair of parallel sides.

questionnaire
D3.1

A questionnaire is a list of questions used to gather information in a survey.

Glossary

radius
S1.6, S2.2, S2.3, S2.4

The radius of a circle is the distance from the centre to the circumference.

random
D2.3, D4.6

A selection is random if each object or number is equally likely to be chosen.

range
D1.2, D1.5, D3.5

The range is the difference between the largest and smallest values in a set of data.

ratio
N1.7, N1.8, A3.4, S3.5,
P1.4, S4.4, S4.5

Ratio compares the size of one part with the size of another part.

raw data
D1.6, D3.2, D3.4, D3.5

Raw data is data that has been collected but not ordered in any way.

rearrange
A5.5

You rearrange a formula by making the subject a different variable, for example, $A = l \times w$ and $w = \dfrac{A}{l}$.

recurring decimal
N1.5, N2.8

A recurring decimal has an unlimited number of digits, which form a repeating pattern, after the decimal point, for example, $\frac{1}{3} = 0.333...$

reflect, reflection
S3.1

A reflection is a transformation in which corresponding points in the object and the image are the same distance from the mirror line.

regular polygon
S1.3

A regular polygon has equal sides and equal angles.

relationship
D3.3

A relationship is a link between objects or numbers.

remainder
N2.7

A remainder is the amount left over when one quantity is divided by another. For example, $9 \div 4 = 2$ remainder 1.

represent
D1.1

You represent data when you display it in the form of a diagram.

rotate, rotation
S3.1

A rotation is a transformation in which every point in the object turns through the same angle relative to a fixed point.

rotational symmetry
P1.5

A shape has rotational symmetry if, when turned, it fits onto itself more than once during a full turn.

round
A3.6, N2.3, N2.6, N2.7

You round a number by expressing it to a given degree of accuracy. For example, 639 is 600 to the nearest 100 and 640 to the nearest 10.
To round to one decimal place means to round to the nearest tenth, for example, 12.47 is 12.5 to 1 dp.

rule
A1.2

A rule describes the link between objects or numbers, for example, the rule linking 2 and 6 might be +4 or ×3.

sample
D1.1, D1.6

A sample is part of a population.

sample space (diagram)
D2.1, D4.1

A sample space diagram records all the outcomes of an experiment.

scale, scale factor
N1.7, N1.8, A3.4, S3.5, S3.6
S4.5, B1.1

A scale gives the ratio between the size of the object and its diagram. A scale factor is the multiplier in an enlargement.

scale drawing
S1.7, S3.6, S4.4

A scale drawing of an object has every part reduced or enlarged by the same amount, the scale factor.

scatter graph
D1.3, D3.3

A scatter graph is a graph on which pairs of observations are plotted.

secondary (data)
D3.1

Data already collected is secondary data.

sector
S1.6, S2.3

Any two radii (plural of radius) will split a circle into two sectors.

segment
S1.6

Any line across a circle will split the circle into two segments.

selling price
N1.5

The price at which an item is sold.

semicircle
S1.6

A diameter splits a circle into two equal halves called semicircles.

sequence
A1.1, A1.2, A2.1, A4.4

A sequence is a set of numbers or diagrams that follow a rule.

similar
S3.5

Two shapes are similar if the angles are the same and corresponding lengths are in proportion.

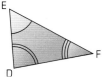

These triangles are similar.
The angles are equal.
DE = 2AB, DF = 2AC, EF = 2BC

simplest form
N1.2

A fraction (or ratio) is in its simplest form when the numerator and denominator (or parts of the ratio) have no common factors.
For example, $\frac{3}{5}$ is expressed in its simplest form.

Glossary

simplify
N1.1, A3.1, A5.2

To simplify a fraction, you divide the numerator and denominator by their highest common factor.
To simplify an expression, you gather all like terms together into a single term.

simulation
D4.4

A simulation is a mathematical model.

sine
B1.2, B1.4, B1.6

The sine of an angle in a right-angled triangle is the ratio of the opposite side o to the hypotenuse h.

$$\sin x = \frac{o}{h}$$

sketch
S1.1

A sketch shows the general shape of a graph or diagram.

skew
D3.4

A distribution is skew if most of the data are at one end of the range.

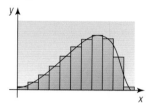 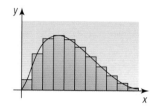

solid (3-D) shape: cube, cuboid, prism, pyramid, square-based pyramid, tetrahedron
S4.3

A solid is a shape formed in three-dimensional space.

cube

six square faces

cuboid

six rectangular faces

prism

the cross section is constant

pyramid

the faces meet at a common vertex

tetrahedron

all the faces are equilateral triangles

square-based pyramid

the base is a square

solution (of an equation)
A3.3, A3.4, A3.6, A5.3

The solution of an equation is the value of the variable that makes the equation true.

solve
A3.2, A5.3

To solve an equation you need to find the value of the variable that will make the equation true.
To solve a problem you need to find the correct answer.

specify
D1.1

You specify a problem by setting it down in detail.

square number, squared
N1.9, A4.1, A4.3

If you multiply a number by itself the result is a square number, for example, 25 is a square number because $5^2 = 5 \times 5 = 25$.

square root
N1.9, A3.3, N2.1

A square root is a number that when multiplied by itself is equal to a given number. For example $\sqrt{25} = 5$, because $5 \times 5 = 25$.

statistics
D1.6

Statistics is the collection, display and analysis of information.

stem-and-leaf diagram
D1.3

A stem-and-leaf diagram is a way of displaying grouped data. For example, the numbers 29, 16, 18, 8, 4, 16, 27, 19, 13 and 15 could be displayed as:

0	4 8
1	3 5 6 6 8 9
2	7 9

Key: | 0 | 4 | means 4

straight-line graph
A4.5

When coordinate points lie in a straight line they form a straight-line graph. It is the graph of a linear equation.

This is the graph of the equation
$y = x + 1$.

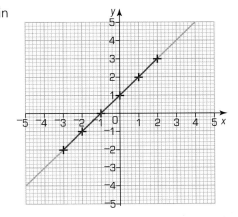

subject
A5.5

The subject of a formula is the term on its own on the left of the equals sign.

substitute
A3.2, A5.1

When you substitute you replace part of an expression with a particular value.

success, successful
D2.3, D2.4

A trial is a success if its result is the required outcome.

sum
S1.3

The sum is the total and is the result of an addition.

surface, surface area
S2.6, S4.6

The surface area of a solid is the total area of its faces.

survey
D3.1

A survey is an investigation to find information.

symmetric, symmetrical
D3.4, D3.5

A shape is symmetrical if a line divides it into two equal parts. The line is a line of symmetry.

systematic
A3.5

A systematic approach to a problem means you start with the simplest case and gradually build up to more complicated ones, one step at a time.

T(n)
A1.1, A1.2, A1.3

T(n) is the notation for the general, nth, term of a sequence. For example, T(3) is the third term.

Glossary

tangent (of an angle)
B1.4, B1.6

The tangent of an angle in a right-angled triangle is the ratio of the opposite side o to the adjacent side a.

$$\tan x = \frac{o}{a}$$

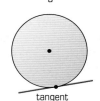

.

tangent (to a circle)
S1.6

A tangent to a circle is a line that touches the circle in one place.

tangent

tenth
N2.3

A tenth is 1 out of 10 or $\frac{1}{10}$.
For example, 0.5 has 5 tenths.

term
A1.1, A1.2, A2.1

A term is a number or object in a sequence or part of an expression.

tessellate, tessellation
S1.2, S1.5

A tessellation is a tiling pattern with no gaps. Shapes will tessellate if they can be put together to make such a pattern.

theoretical probability
D4.4, D4.5, D4.6

A theoretical probability is worked out without an experiment by considering all the possible outcomes.

term-to-term rule
A1.1, A1.3

A term-to-term rule links a term in a sequence to the previous term.

thousandth
N2.3

A thousandth is 1 out of 1000 or $\frac{1}{1000}$.
For example, 0.002 has 2 thousandths.

three-dimensional (3-D)
S3.4

Any solid shape is three-dimensional.

time axis
A2.2

The time axis is the horizontal axis on a distance–time graph.

transformation
S3.3

A transformation moves a shape from one place to another.

translate, translation
S3.1

A translation is a transformation in which every point in an object moves the same distance and direction. It is a sliding movement.

tree diagram
D4.3, D4.4

A tree diagram shows all the possible outcomes of one or more events.
This tree diagram shows the possible outcomes of flipping a coin twice.

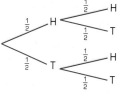

You write the outcomes at the end of branches and the probabilities on the branches.

trend
A4.8

A trend is a relationship between observed data and an independent variable such as time.

trial
D2.2, D2.4

A trial is a single observation in an experiment.

trial and improvement
A3.5, P1.3, P1.4

To find the answer to a complex calculation it is sometimes easier to estimate the answer then improve the estimate.
This is called trial and improvement.

triangle: equilateral, isosceles, scalene, right-angled
S1.5

A triangle is a polygon with three sides.

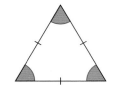

equilateral

three equal sides

isosceles

two equal sides

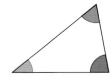

scalene

no equal sides

right-angled

one angle is 90°

triangular number
A1.1

The triangular numbers form the sequence 1, 3, 6, 10, 15, 21, 28, ...
They are the number of dots in a triangular pattern.

two-way table
D3.2

A two-way table links two independent variables, for example, the result when you toss two dice and add the scores.

		Dice 1					
		1	2	3	4	5	6
Dice 2	1	2	3	4	5	6	7
	2	3	4	5	6	7	8
	3	4	5	6	7	8	9
	4	5	6	7	8	9	10
	5	6	7	8	9	10	11
	6	7	8	9	10	11	12

unit fraction
N1.2, N1.3

A unit fraction has 1 as the numerator, for example, $\frac{1}{2}, \frac{1}{7}, \frac{1}{23}$.

unitary method
N1.4, N1.6, S3.6

In a unitary method you first work out the size of a single unit and then scale it up or down.

unknown
A5.3

An unknown is a variable. You can often find the value of an unknown by solving an equation.

value
P1.6

The value is the amount an expression or variable is worth.

variable
A5.1, D3.3

A variable is a symbol that can take a range of values.

vector
S3.1

You can use a vector to specify a translation. For example, $\binom{3}{4}$ means you move the object 3 units to the right and 4 units up.

To move left or down you use negative numbers in the vector.

vertex, vertices
S1.2

A vertex of a solid is a point at which two or more edges meet.
A vertex of a 2-D shape is where two sides meet.

vertex

vertical
S4.3, A4.8, A4.9, B1.1

Vertical means straight up and down, at right angles to the . horizontal.

vertically opposite angles
S1.4, S4.2

When two straight lines cross they form two pairs of equal angles called vertically opposite angles.

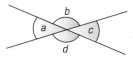

$a = c$ $b = d$

volume: cubic millimetre, cubic centimetre, cubic metre
S2.5, S4.6

The volume of an object is a measure of how much space it occupies.

A1 Check in

1 a 13, 16
  b 3, 6
  c ⁻9, ⁻13
  d 2.7, 2.9, 3.1
2 a 7
  b 14
  c 4
  d 1

A1 Check out

1 a 9, 13, 19
  b ⁻3, 7, 22
  c 8, 4, ⁻2
  d ⁻17, ⁻11, ⁻2
  e 0.5, ⁻4.5, ⁻12
2 a Start at 4 and add 3; $3n + 1$
  b Start at 9 and add 2; $2n + 7$
  c Start at 20 and subtract 2; $22 - 2n$
  d Start at 8 and add 5; $5n + 3$
3 a 6, 8, 10, 12
  b $T(10) = 24$, $T(25) = 54$
  c $6 + (24 \times 2)$
  d $T(n) = 2(n - 1) + 6 = 2n + 4$

A2 Check in

1 $3n + 4$

2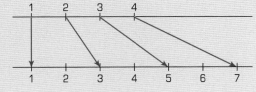

3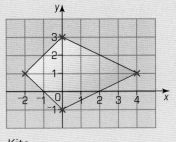

Kite

A2 Check out

1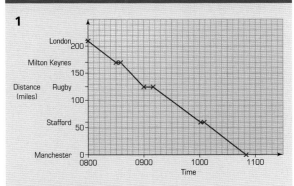

2 a 100 miles
  b 40 miles, 30 minutes
  c 80 mph

1 HCF = 4, LCM = 120

2 $\frac{1}{2} = \frac{12}{24}$, $\frac{24}{40} = \frac{6}{10}$, $\frac{2}{3} = \frac{24}{36}$, $\frac{15}{60} = \frac{3}{12}$

3 a $\frac{1}{5}$
  b 45%
  c 0.15
  d 44%
4 a 1 : 3
  b 10 : 3
  c 3 : 5
  d 200 : 1

1 a $\frac{23}{24}$
  b i $\frac{3}{10}$
    ii

  c 22
  d $4\frac{2}{5}$
2 a 238 g   b 17.8%
3 a i Smaller   ii Larger
    iii Larger   iv Smaller
  b i $10 \times \frac{1}{2} = 5$ ii $10 \div \frac{1}{2} = 20$
    iii $10 \times 1\frac{1}{2} = 30$ iv $10 \div 2 = 5$
  c i 4   ii 25   iii 35   iv $7\frac{1}{7}$
4 a $200 \text{ g} \times \frac{300}{200} = 300$ g, $\frac{40}{200} = 20\%$
  b Students' own answers
5 a The answer should be less than £340.
  b £289
  c John divided by $\frac{85}{100}$ when he should have multiplied.

1 a ^-y
  b $p - r + 7$
2 a $p = 5$
  b $n = 2$
3 $y = {}^-5, {}^-3, {}^-1, 1, 3$
4 a $2t - 6y$
  b $24 - 3x$

1 a i $x = 10$
    ii $x = 2.5$
  b $11y = 220$, $y = 20$
  c $x = 12.5$ cm
2 a

Estimate	$2x^2$	$3x$	$2x^2 - 3x$	Too small / too big
8	128	24	104	too small
8.1	131.22	24.3	106.92	correct

  b $6x^2 + x = 84$, $x = 3.7$ to 1 dp

1 $a = 25°$, $b = 120°$

2 a Isosceles triangle: two equal sides, two equal angles.

b Parallelogram: opposite sides parallel and equal, opposite angles equal.

1 a 30° because $x + 60° + 90° = 180°$

b 60° because $y + 30° + 90° = 180°$

c 10 cm because triangle ABC is equilateral.

d 10 cm because triangle ABC is equilateral.

e Gill is correct. Paul should have used the vertical height, which will be less than 10 cm, and halved the result.

2 No, the total of the two shorter sides needs to be greater than the length of the longest side.

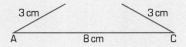

1 a 1, 2, 3, 6, 7, 7, 8

b 0.6, 0.8, 1.2, 1.2, 1.3, 1.4, 1.5, 1.9

2

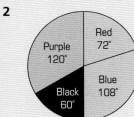

Shows proportions clearly.

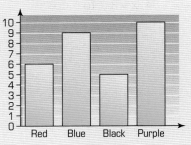

Shows actual values.

1 Data you collect yourself is primary data for example, a survey of classmates. Data that is already available is secondary data, for example, nationally gathered statistics on the internet.

2 a Small sample size, possibly mostly boys, holiday timing so more TV watched. Primary data so more accurate, large variety of detail available.

b Expensive, time-consuming, too detailed. Larger sample size so possibly more accurate, more detailed results.

3 a Questions that don't work as expected can be changed as a result of the pilot survey. New questions could also be added.

b Sample survey question with range of options.

4 a 3-D charts look attractive. However, they can be misleading.

b Scales that don't start at zero, for example, can be used to make results look more dramatic than they are.

S2 Check in

1 Perimeter = 28 m, Area = 28 m^2

2 a 1.2 m
  b 26 100 m
  c 3.7 cm
  d 1920 cm

3 a 20 cm^2
  b 10 cm^2

4 a 15 m^3
  b 6.3 cm^3

S2 Check out

1 a 13 people
  b 16 people
  c 17 people

2 57°

3 No, they would have less than 38 cm of space each.

N2 Check in

1 a $^-7$
  b $^-3$
  c $^-10$
  d 10
  e 1

2 $\frac{18}{36}$, $\frac{24}{36}$, $\frac{27}{36}$, $\frac{30}{36}$; $\frac{5}{6}$ is largest

3 a 1114
  b 1719
  c 9594
  d 23

N2 Check out

1 a £1200
  b £21

2 a $4 \times (3.9 \div 1.3) = 12$
  b $100 \times 0.8 \div 5 = 16$

3 24.457 m

4 a £12.85
  b 230 bags

1 **a** 3^4
  b m^4
  c f^7
2 **a** $2 \times 2 \times 5$
  b $2 \times 2 \times 5 \times 3$
  c $2 \times 2 \times 2 \times 2 \times 2$
  d $2 \times 2 \times 5 \times 5$
3 $^-11, \ ^-3, \ 1, \ 5$

1 **a i** A: gradient 2, intercept $^+3$
       B: gradient 2, intercept $^-2$
    ii A: $y = 2x + 3$
       B: $y = 2x - 2$
  b i A: gradient 3, intercept $^-2$
       B: gradient 1, intercept $^-2$
    ii $(0, \ ^-2)$

2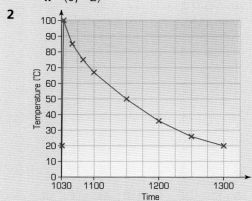

3 **a** 1220 **b** 1055
  c 1120 **d** 2 h 28 min

1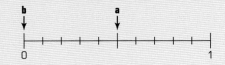

2 C, R, O, D, I, L, E

3 **a** 75%
  b 40%
  c 35%
  d 35%

1 $p(\text{orange}) = 1 - p(\text{red}) - p(\text{green})$
  $= 1 - 0.2 - 0.35$
  $= 0.45$

2 Game A:
  Chance of winning $= \frac{3}{36} = \frac{1}{12}$

  Game B:
  Chance of winning $= \frac{6}{36} = \frac{1}{6}$

  There is a greater chance of winning on Game B.

1

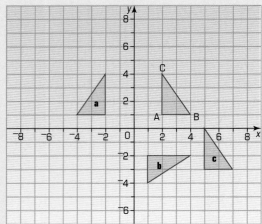

2 **a** Cuboid

   b Tetrahedron

   c Pentagonal prism

3 **a** 1 : 4   **b** 2 : 5

   c 5 : 8   **d** 6 : 23

1 **a** and **b**

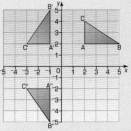

   c Rotation through 180° about (0, 0)

2 **a–d**

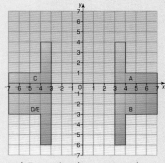

   e D and E are in the same place.

3 No. Any hexagon drawn on four mirror lines ends up with six lines of symmetry.

1 96 cm by 72 cm

2 The longer side is (12 – p) cm, so the area is $p(12 - p) = 12p - p^2$.

3 275 ml

4 The man walks at a steady pace to the park, which is 2 km away. He then stays at the park for about an hour. He walks most of the way home very quickly, and the last 0.5 km very slowly.

5 $x = 55°$ (angles on a straight line)

   $y = 30°$ (vertically opposite angles)

   $z = 95°$ (angles in a triangle sum to 180°)

A5 Check out

1 **a** $18 - 2x + (3y - 7) = 0$

  $18 - 2x + (6x - 5) = 23$

  $x = 2.5, y = {}^-2$

  b $^-10.5$

2 **a** $l = \frac{2V}{bh}$

  b 75 cm

3 **a**

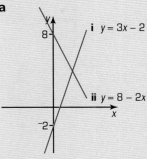

i $y = 3x - 2$

ii $y = 8 - 2x$

  b (2, 4)

  c $y = 3x - 2$

1 For example:

Time (t seconds)	Frequency
$16 \leqslant t < 17$	1
$17 \leqslant t < 18$	3
$18 \leqslant t < 19$	3
$19 \leqslant t < 20$	3
$20 \leqslant t < 21$	1
$21 \leqslant t < 22$	1

2 Mean = 18.6 s, median = 18.4 s, modal class = $17 \leqslant t < 18$; $\leqslant 18t < 19$; $19 \leqslant t < 20$; (mode = 19.2 s)

3 For example:

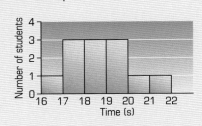

1 **a** For example:

  Grouped frequency table with classes:

  $x < 50$, $50 \leqslant x < 60$, $60 \leqslant x < 70$,

  $70 \leqslant x < 80$, $80 \leqslant x < 90$, $90 \leqslant x < 100$

  or Bar chart for grouped data, above.

  or Stem and Leaf diagram.

  Because data is discrete, need to show low value of 32.

  c Range is large: 67. Most marks are in range 75–99. One mark is very low (32). Distribution is skewed.

2 **b** Time series for oak tree

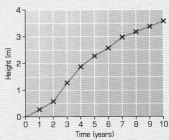

  After 7 years the tree grew more slowly.

1 a 3 equal sides, 3 equal angles, 3 lines of symmetry, rotational symmetry of order 3

b Opposite sides are equal and parallel, no lines of symmetry, rotational symmetry of order 2

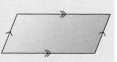

c Adjacent sides are equal, one pair of opposite angles are equal, 1 line of symmetry, no rotational symmetry (order 1)

2 For example:

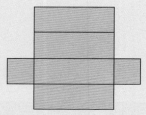

3 a 5 : 1
 b 3 : 1
 c 1 : 6000
 d 3 : 16

1 a 380 cm^2
 b 33.8 cm

2 a Accurate drawing of triangle:

b 121 m

3 a

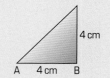

b

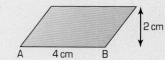

c

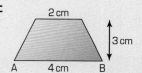

d

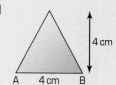

1 a $\frac{1}{13}$
  b $\frac{1}{3}$
2 a 30
  b 21
  c 12

1 a

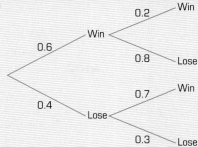

b 0.76
c 0.12
2 a $\frac{1}{16}$, $\frac{1}{8}$, $\frac{3}{16}$, $\frac{1}{4}$, $\frac{3}{16}$, $\frac{1}{8}$, $\frac{1}{16}$
  b $\frac{3}{50}$, $\frac{7}{50}$, $\frac{4}{25}$, $\frac{6}{25}$, $\frac{9}{50}$, $\frac{3}{20}$, $\frac{2}{25}$
  c When you convert the fractions to decimals the probabilities are similar. The dice seem to be fair.

1 3
2 a 9   b 144   c 38.44   d 302.76
  e 9   f 14   g 17.1   h 6.8
3

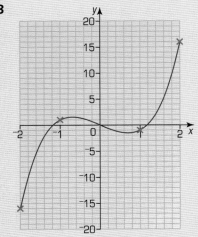

It is a cubic graph.
or: It is a smooth curve.

1 a $6^2 + 7^2 = 85$ but $8^2 = 64$, so the triangle cannot be right-angled by Pythagoras' theorem.
  b 1.73 m
  c 26 units
2 a $x = 29.0°$
  b $y = 6.11$ km
  c $z = 8.89$ m
3 a $s^2 = 3^2 + 3^2 = 18$ by Pythagoras' theorem, so $s = \sqrt{18} = 4.242\,640\,687$ m
  b The answer should be rounded to 2 dp, giving $s = 4.24$ m.

Index

A

acceleration, 134
accuracy, 112, 172, 214
addition
    fractions, 18–19
    mental arithmetic, 106–7
adjacent sides, 236–7, 238, 240
algebra, equation solving, 46
algebraic expressions, 172
algebraic fractions, 182–3
alternate angles, 58, 206
angle bisectors, 68
angle sums, 54
angles
    alternate, 58, 206
    corresponding, 58, 150, 206
    exterior, 54, 56, 60
    interior, 54, 56, 60
    and lines, 58–9
    in polygons, 54–7
    properties, 206–7
    in quadrilaterals, 60
    in scale drawings, 210
    straight lines, 170
    triangles, 170, 206
    vertically opposite, 58, 206
anomalies, 78
approximations, 110, 112, 168, 172
arcs, 62, 232
area
    circles, 90–1, 92
    measuring, 86–7
    metric units, 86, 94
    surface, 96, 214–15
arithmagons, 180
average speed, 12, 134
averages, 74, 196, 198
axes, graphs, 12, 82

B

balance method, 40, 176
bar charts, 76
bases, 122
best estimate, 48–9
bias, 144
bisectors
    angle, 68
    perpendicular, 66
brackets, 34, 42, 178–9
    calculators, 114
    expanding, 178
    multiplying, 178, 180

C

calculator operations, 34, 114–15, 234
calculators
    power key, 100
    scientific, 114
Celsius, 184
centimetres, 86, 94
centres of enlargement, 156
centres of rotation, 148
charts
    bar, 76
    doughnut, 78
    interpreting, 78–9
    pie, 76, 78
    three-dimensional, 82
    see also diagrams; graphs
chords, 62
circles
    area, 90–1, 92
    circumference, 62, 88–9, 90, 92
    formulae, 92
    loci, 66
    problems, 92–3
    properties, 62–3
    triangles in, 232–3
circumference, 62, 88–9, 90, 92
combined events, 218–19, 222, 228
common denominators, 18, 106, 120,
    142, 182
common factors, 108, 178
    fractions, 20
compensation, 106
conclusions, 200
congruence, 150, 204
congruent triangles, 150, 204
construction lines, 64
constructions, geometrical, 51–70
continuous data, 74, 76, 192
conventions, 52, 150
conversions, 32, 210
coordinate pairs, 126
correlation, 194–5
corresponding angles, 58, 150, 206
corresponding sides, 150
cosines, 234–7, 238, 240, 242
cross-sections, 154
cube roots, 34, 100
cubes, volume, 94
cubic centimetres, 94
cubic kilometres, 94
cubic metres, 94
cubic millimetres, 94
cuboids, 154

surface area, 96
volume, 94, 214
cylinders, 96

D

data
    continuous, 74, 76, 192
    grouped discrete, 192
    handling, 71–84
    interpretation, 72, 82
    organising, 192–3
    primary, 72, 190
    raw, 82, 192, 196
    representation, 72, 76
    secondary, 72, 190
    skewed, 74
    tables, 190, 192
data collection, 72, 190–1
data collection sheets, 190
decimal places, 48, 104
decimal system, 102
decimals
    and fractions, 108
    multiplying, 110
    and percentages, 108
    recurring, 26, 114
    rounding, 104
definitions, 52
degrees (angular), 52, 234
degrees (temperature), 184
denominators, 18, 112
    common, 18, 106, 120, 142, 182
derived properties, 52, 54
diagrams
    frequency, 76
    sample space, 138, 218
    stem-and-leaf, 76
    tree, 222–5, 226
    see also charts; graphs
diameter, 62, 88, 90
dice, biased, 144
dimensions, 94, 210
direct proportion, 44
distance, 162
distance–time graphs, 12–13, 134–5
distributions see frequency distributions
dividends, 112
division, 112–13
    calculators, 114
    fractions, 22–3
    indices, 100, 122
    integers, 22
    mental arithmetic, 108–9

powers of ten, 102–3
doughnut charts, 78

E

elevations, 208
enlargements, 44, 156–7, 234
equally likely outcomes, 138
equations, 175–88
    letter symbols, 38
    non-linear, 42
    solving, 37–50, 180–1
    graphically, 186–7
    of straight lines, 44, 126, 128
    see also linear equations
equidistant points, 62
equilateral triangles, 52, 56
equivalent expressions, 38
equivalent fractions, 18, 142, 182
estimation, 46, 48–9, 104, 110
euros, 32
events
    combined, 218–19, 222, 228
    independent, 218
    probability of, 138–9
    random, 228
exceptions, 78
expected frequencies, 140, 226
experimental probability, 144–5,
    217–30
experiments, 140
    predictions, 226
    probability, 228–9
    simulating, 224
expressions
    algebraic, 172
    equivalent, 38
    simplifying, 178
    substituting, 40
    terms, 178
exterior angles, 54, 56, 60
extreme values, 74, 196

F

factor pairs, 118
factor trees, 118
factorisation, 178
factors, 108
    common, 20, 108, 178
    finding, 118–19
    prime, 18, 118, 120–1
    scale, 30, 32, 44, 156, 234
Fahrenheit, 184
favourable outcomes, 138, 220
formulae, 175–88
    circles, 92
    letter symbols, 38
    rearranging, 184

subjects of, 184
transformations, 184–5
triangles, 242
fractions
    adding, 18–19
    algebraic, 182–3
    calculations, 24
    common factors, 20
    and decimals, 108
    dividing, 22–3
    equivalent, 18, 142, 182
    improper, 20
    multiplying, 20–1
    and percentages, 26, 108
    simplifying, 18, 108
    subtracting, 18–19
frequencies, expected, 140, 226
frequency diagrams, 76
frequency distributions
    comparing, 80–1
    shape, 198–9
    skewed, 196
    symmetric, 196
frequency tables, 192
front elevation, 208
function machines, 10–11
functions, 9–16, 175–88
    graphs of, 117–36, 186
    implicit, 130–1
    inverse, 10
    letter symbols, 38
    linear, 126, 128, 130

G

general terms, 2, 4, 6, 10
geometrical constructions, 51–70
geometrical language, 52–3
geometrical reasoning, 51–70, 170–1,
    203–16
gradients, 126, 128–9, 186
graphs, 9–16, 76
    axes, 12, 82
    distance–time, 12–13, 134–5
    equation solving, 186–7
    of functions, 117–36, 186
    interpreting, 14–15, 78–9, 132–5
    line, 76
    of linear functions, 126, 128, 130
    real-life, 132–3
    scatter, 76, 78, 194–5
    see also charts; diagrams; linear
    graphs
grouped discrete data, 192

H

halfway values, 48
hectares, 86

hexagonal prisms, 96
hexagons, angles, 56
highest common factor (HCF), 18,
    120
hundredths, 102
hypotenuse, 64, 232, 234, 236, 238,
    240
hypotheses, 72, 190, 200

I

identities, 176–8
    letter symbols, 38
images, 148, 152
implicit functions, 130–1
improper fractions, 20
independent events, 218
index form, 100
index notation, 100, 122–3
indices, 100–3
    dividing, 100, 122
    laws, 122
    multiplying, 100, 122
    negative, 100
    notation, 100, 122–3
    order of operations, 34
    positive, 100
    powers of ten, 102–3
    problem solving, 124–5
inequalities, 104, 192
inferences, 78
integers, 102, 112
    dividing, 22
    multiplying, 20
intercepts, 126, 128, 130, 186
interior angles, 54, 56, 60
interpretation
    data, 72, 82
    graphs, 14–15, 78–9, 132–5
intersections, 186
inverse functions, 10
inverse operations, 10
inverse relationships, 10, 22, 28
isometric paper, 208
isosceles triangles, 60

K

kilometres, 86, 94
kites, 60, 204, 206, 208, 210

L

length, metric units, 94
less than, 192
letter symbols, 38–9
line graphs, 76
line segments, 156
linear equations
    balancing, 40–1

Index

solving, 40, 180
linear functions, graphs of, 126, 128, 130
linear graphs, 44
 gradients, 126, 128–9, 186
 intercepts, 126, 128, 130, 186
 plotting, 126–7
linear sequences, 6
lines
 and angles, 58–9
 construction, 64
 mirror, 148
 parallel, 58, 206
 vertical, 208
 see also straight lines
lines of best fit, 76
lines of symmetry, 60, 204
loci, 66–9
lowest common multiple (LCM), 18, 120, 182

M

mapping diagrams, 10
maps, 158, 212–13
mean, 74, 196
median, 74, 196
mental arithmetic, 106–9
metres, 86, 94
metric units, 86, 94
 conversions, 86
millimetres, 86, 94
mirror lines, 148
modal class, 74, 80, 196
mode, 74, 196
multiplication, 110–11
 brackets, 178, 180
 calculators, 114
 decimals, 110
 fractions, 20–1
 indices, 100, 122
 integers, 20
 mental arithmetic, 108–9
 negative numbers, 42
 powers of ten, 102–3
multiplicative inverse, 22, 28
mutually exclusive outcomes, 140–1, 142, 218, 220–1

N

negative correlation, 194
negative numbers, 34
 multiplying, 42
negative terms, 178
nets, 208, 214
non-linear equations, 42–3
number calculations, 99–116

numbers
 negative, 34, 42
 positive, 42
 prime, 118
 square, 118
numerators, 20, 142

O

operations
 calculator, 34, 114–15, 234
 inverse, 10
 order of, 34–5
opposite sides, 236–7, 238, 240
outcomes, 138, 218, 222, 228
 equally likely, 138
 favourable, 138, 220
 mutually exclusive, 140–1, 142, 218, 220–1

P

parallel lines, 206
 angles, 58
parallelograms, 52, 60
 area, 86
partitioning, 106, 108
patterns, and sequences, 2, 4, 10
percentages
 changes, 24, 26, 28–9
 and decimals, 108
 decrease, 24, 26
 and fractions, 26, 108
 increase, 24, 26
 and proportions, 24
perimeters, 88, 166
perpendicular bisectors, 66
perpendiculars, constructing, 64
pi (π), 88, 90
pie charts, 76, 78
place value, decimals, 48, 104
planes of symmetry, 154–5
plans, 208
polygons, 52
 angles, 54–7
population pyramids, 80
position-to-term rule, 2, 6
positive correlation, 194
positive numbers, 42
pounds (money), 32
powers *see* indices
predictions, 226
primary data, 72, 190
prime factors, 18, 118, 120–1
prime numbers, 118
prisms, 96, 154
 surface area, 96, 214
 volume, 96, 214–15

probability, 137–46
 calculating, 142–3, 164
 experimental, 144–5, 217–30
 simulations, 224
 theoretical, 138, 144, 217–30
 see also events
probability experiments, 228–9
probability scales, 138
problem solving, 161–74
 checking results, 172–3
 and indices, 124–5
 strategies, 166–7
 triangles, 242–3
projections, 208
proportional problems, solving, 44–5
proportional reasoning, 17–36, 168–9
proportionality, 32–3
proportions, 158–9
 of amounts, 23–5
 comparing, 26–7
 direct, 44
 and percentages, 24
 and ratios, 32
protractors, 52
pyramids, 154
 population, 80
Pythagoras' theorem, 240–1, 242

Q

quadratic sequences, 6–7
quadrilaterals, 54
 angles, 60, 204

R

radius, 62, 88, 90
random events, 228
range, 74, 80, 198
ratios, 30–1, 44
 and maps, 212–13
 and proportion, 32
 and scale drawings, 210–11
 unitary, 30, 210
raw data, 82, 192, 196
reasoning, proportional, 17–36, 168–9
rectangular prisms, 96
recurring decimals, 26, 114
reflections, 148
regular polygons, angles, 56–7
reports, statistical, 189–202
representations, 72, 76
 two-dimensional, 208
results, checking, 172–3
rhombuses, 60
right-angled triangles, 204, 232
 constructing, 64–5
 horizontal bases, 234, 236

hypotenuse, 232, 234, 236, 238, 240
  Pythagoras' theorem, 240–1, 242
  trigonometry, 232–9
  vertical height, 234, 236
roots, 100–1
  cube, 34, 100
  square, 34, 42, 100
rotations, 148
rounding, 104–5, 106, 108, 112

S

sample size, 82
sample space diagrams, 138, 218
scale drawings, 64, 158, 210–11
scale factor method, 32
scale factors, 30, 32, 44, 156, 234
scales, 30, 64, 158–9, 232
  maps, 212
  probability, 138
scaling method, 32
scatter graphs, 76, 78, 194–5
scientific calculators, 114
secondary data, 72, 190
sectors, 62
segments, 62
  line, 156
semicircles, 62
sequences, 1–8, 124
  linear, 6
  *n*th term, 2, 4, 6
  and patterns, 2, 4, 10
  practical contexts, 4–5
  quadratic, 6–7
  rules, 2, 6
  terms, 2
shapes
  congruence, 150, 204
  enlargements, 156–7
  similarity, 156–7
  solving problems in, 60–1
  three-dimensional, 154, 214
  two-dimensional, 150, 204–5
side elevation, 208
similarity, 156–7
simulations, probability, 224
sines, 234–7, 238, 240, 242
skewed data, 74
skewed distributions, 196
slopes (gradients), 126, 128–9, 186
solids, 208
speed, average, 12, 134
square centimetres, 86, 94
square kilometres, 86, 94

square metres, 86, 94
square millimetres, 86, 94
square numbers, 118
square roots, 34, 42, 100
squares, 52
  area, 86, 94
statistical enquiries, stages, 72
statistical reports, 189–202
statistics
  calculating, 74–5
  choosing, 196–7
  misleading, 82–3
stem-and-leaf diagrams, 76
straight lines
  angle sum, 54
  angles, 170, 206
  equations of, 44, 126, 128
straight-line graphs *see* linear graphs
substitution, 172, 180
  in expressions, 40
subtraction
  fractions, 18–19
  mental arithmetic, 106–7
surface area
  cuboids, 96
  prisms, 96, 214
surveys, 72
  data collection, 190
symmetric distributions, 196
symmetry
  lines of, 60, 204
  planes of, 154–5

T

tables
  data, 190, 192
  frequency, 192
  graph plotting, 186
  two-way, 192
tangents, 62, 238–9, 242
temperature
  graphs, 14
  units, 184
tenths, 102
terms
  expressions, 178
  general, 2, 4, 6, 10
  negative, 178
  *n*th, 2, 4, 6
  in sequences, 2
term-to-term rule, 2, 6
theoretical probability, 138, 144,
  217–30

thousandths, 104
three-dimensional charts, 82
three-dimensional objects, two-
  dimensional representations, 208–9
three-dimensional shapes, 154
  surface area, 214
transformations, 147–60
  combining, 152–3
  describing, 150–1
  of formulae, 184–5
translations, 148
trapeziums, area, 86
tree diagrams, 222–5, 226
trends, 132
  exceptions, 78
trial and improvement method, 46–7,
  100, 166, 170
trials, 140, 142, 144, 224, 226
triangles
  angle sum, 54
  angles, 170, 206
  area, 86
  in circles, 232–3
  congruent, 150, 204
  equilateral, 52, 56
  formulae, 242
  isosceles, 60
  problem solving, 242–3
  see also right-angled triangles
triangular prisms, 96
trigonometry, 231–44
two-dimensional shapes, 150
  properties, 204–5
two-way tables, 192

U

unitary method, 24, 28, 32, 158
unitary ratios, 30, 210
units
  metric, 86, 94
  temperature, 184

V

variables, 176, 194
vectors, 148
vertical lines, 208
vertically opposite angles, 58, 206
vertices, 54
volume
  cuboids, 94, 214
  measuring, 94–5
  metric units, 94
  prisms, 96, 214–15